EVANSTON PUBLIC LIBRARY
P9-CRE-739

written in
WATER

written in WATER

MESSAGES OF HOPE FOR EARTH'S MOST PRECIOUS RESOURCE

Edited by Irena Salina

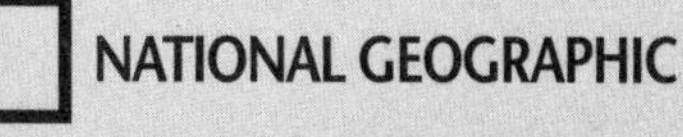

WASHINGTON, D.C.

Published by the National Geographic Society
1145 17th Street N.W., Washington, D.C. 20036

ISBN: 978-1-4262-0572-9

Library of Congress Cataloging-in-Publication Data
Written in water : messages of hope for earth's most precious resource/ Irena Salina, [editor].
p. cm.
Includes bibliographical references.
ISBN 978-1-4262-0572-9 (alk. paper)
1. Water--Environmental aspects. 2. Aquatic ecology. I. Salina, Irena, 1968-
GB662.3.S25 2009
333.91--dc22
2009046141

The National Geographic Society is one of the world's largest nonprofit scientific and educational organizations. Founded in 1888 to "increase and diffuse geographic knowledge," the Society works to inspire people to care about the planet. It reaches more than 325 million people worldwide each month through its official journal, *National Geographic,* and other magazines; National Geographic Channel; television documentaries; music; radio; films; books; DVDs; maps; exhibitions; school publishing programs; interactive media; and merchandise. National Geographic has funded more than 9,000 scientific research, conservation and exploration projects and supports an education program combating geographic illiteracy.

For more information, please call 1-800-NGS LINE (647-5463) or write to the following address:

National Geographic Society
1145 17th Street N.W.
Washington, D.C. 20036-4688 U.S.A.

Visit us online at www.nationalgeographic.com

For information about special discounts for bulk purchases, please contact
National Geographic Books Special Sales: ngspecsales@ngs.org

For rights or permissions inquiries, please contact National Geographic Books Subsidiary Rights: ngbookrights@ngs.org

Interior design: Cameron Zotter

09/WOR/1

CONTENTS

FOREWORD 7
Peter Gleick

THE END OF THE COLORADO 13
Fishing with the Cucapá | Frank Clifford

A RIVER STORY 26
Cleaning Up the Nashua | Marion Stoddart

LIQUID ASSETS ON STEEP SLOPES 36
Solving Water Shortages Through Ancient Knowledge | Anupam Mishra

HONEST HOPE 46
Searching for a New Water Ethic | Sandra Postel

RESPECT FOR WATER 60
Reflecting on a Lifetime of Environmental Reporting | Fred Pearce

ANOTHER AFRICA IS POSSIBLE 76
Witnessing the Consequences of Kenya's Flower Farms | Maude Barlow

THE UNMENTIONABLES 82
Talking About Sanitation | Rose George

BRIDES OF THE WELL 92
A Short Story | Shekhar Kapur

A WORLD THAT WORKS 101
Breaking Places | Bill McKibben

PHOTOGRAPHY JOURNEY 107
Taking the Long View at Mount Everest | Alton C. Byers

WHAT'S IN A NAME 120
Monitoring the "Miner's Canary" in the Congo | Melanie Stiassny

THE FEAR OF DOING WITHOUT 130
Bringing Fresh Water and New Ideas to the Dominican Republic | Ellsworth Havens

WATER MAN 137
Reviving Arid Villages | Rajendra Singh

A RIDE FOR NATURE 143
Traveling the West by Horseback | William "Waterway" Marks

WATER CHANGES EVERYTHING 162
Building a Charity | Scott Harrison

SAVE WHAT'S LEFT 177
Teaching Kids to Love Rain Forests | Lynne Cherry

TO SAVE A GULL 186
Taxing Yourself | Alex Matthiessen

DYNAMIC FOR CHANGE 205
Recognizing the Importance of the Watershed | Christine Todd Whitman

THE PÁRAMOS OF AYABACA 210
Defending Water and Place | Jorge Recharte

TENDING THE LAND 219
Reclaiming Our Food | Frederick Kirschenmann

A WALK ALONG THE RIVER 236
Caring for Watersheds | Dave Rosgen

ON THE HEADS OF WOMEN 250
Safeguarding Water | Kathy Robb

TWO NOBLE TRUTHS 258
The Ultraviolet Solution to Clean Water | Ashok Gadgil

NATURE ABHORS A MONOPOLY 268
How We, Like Kalahari Bushmen, Can Trade Our Human Right to Water | James Workman

WATER IS LIFE 283
Stories from a Blue Planet | Alexandra Cousteau

How This Book Came to Be **292**, Authors' Websites **294**,
Get Involved **296**, Further Reading **302**

FOREWORD

Peter Gleick

Peter Gleick is co-founder and president of the Pacific Institute for Studies in Development, Environment, and Security in Oakland, California; recipient of the MacArthur Fellows "Genius Award"; and an internationally renowned water visionary.

I'VE WORKED PROFESSIONALLY IN WATER for a long time—beginning in the 1970s. But like many of the authors in this collection, my connection to water goes back to long before my professional career, to childhood expeditions exploring the waters of the Pine Barrens of New Jersey, the northern peninsula of Michigan, the inland lakes of Maine. I watched my buttoned-down, pale, officebound lawyer father go off on a two-week river trip riding wooden dories down the Grand Canyon and come back home a bearded, tanned, happy river god. He told me stories of riding the same river and seeing the same ancient canyon walls seen by an earlier one-armed river god, John Wesley Powell, whose own story inspired so many others to work on water. I've now been down the same stretches of the Colorado River, knowing that decades before my father slept on the same riverbanks, saw those same etched rocks, felt the same sense of awe and wonder over the power of a river. If we protect the river, perhaps my children and their children and more generations will experience that same wonder and awe.

A little bit of oxygen and two little bits of hydrogen. Two of the most basic elements in the universe. Yet they combine to form a magical substance fundamental to life itself: water. We live on a watery planet, unique in our solar system and probably extraordinarily rare in the galaxy. Some have described Earth as the Goldilocks planet. Not too hot, or liquid water couldn't exist, and not too cold, or it would be a frozen wasteland. Just right. And just right it is, with a remarkable combination of water flowing in different forms through different stocks: vast oceans, ice caps, rivers, lakes, water vapor in the atmosphere, and the water that makes up the blood in our veins.

Just as the geological history of the planet can be written in the way the world's water has ebbed and flowed among these stocks, so too the history of humanity can be written in the different "Ages of Water" and the way civilizations have understood, managed, and used water. And just as the history of humanity can be written in water, so too can the history of individuals, as the diverse, poetical, and often intensely personal essays in this book show. The power of water has etched the canyons of the planet and the canyons of our souls.

As humans emerged from the fog of evolution, our relationship to water has evolved. The First Age of Water saw simple and serendipitous uses of water. Hunter-gatherer tribes took water from wherever it was available for drinking and cleaning and used flowing water to remove their wastes, where nature's biological processes recycled them. If water was bad, or unavailable, people moved or got sick and died, but life was short and brutish already.

The development of agriculture and the establishment of permanent communities began to change our relationship to water, ushering in the Second Age of Water. This age saw the first intentional manipulations of the hydrological cycle to move water from places of abundance to places of scarcity, to control floods and droughts, to separate water supply from wastes, and to treat contaminated water. These technological manipulations have become ever more sophisticated over time, as our engineering, chemical, and physical capabilities, knowledge, and expertise have grown. From the first small earthen or rock dams built in ancient Egypt, India, and Mesopotamia, we now seek to control the largest rivers in the world with huge concrete-and-steel infrastructure. From simple sand filters and the use of chemicals to clean and purify wastewaters, we now have the ability to turn the worst quality wastewaters into the purest drinking water using microscopic membranes, ultraviolet light, ozonation, and other techniques. From simple flooding of agricultural fields using rudimentary canals and hand-dug ditches, we now can accurately measure crop water needs and soil moisture with satellites and neutron probes and apply precise amounts of water to specific areas where and when we want.

This Second Age of Water brought enormous benefits as human populations expanded from the millions to the billions. Rampant water-related diseases have been eradicated in the richer countries. Floods and drought that used to kill hundreds of thousands, even millions, of people at a time are increasingly rare. The massive expansion of irrigation that helped drive the green revolution has reduced the vulnerability of the poorest populations to famine and starvation.

But the Second Age of Water is coming to an end, and the globe is still faced with major, unresolved water challenges. The skills, tools, and approaches developed over the past few centuries to manage water must be expanded and applied more widely, since they do not seem to be able to eliminate our water problems alone. It is now the 21st century, but a billion people still lack safe drinking water. Two and a half billion still lack adequate sanitation that keeps human wastes out of their immediate environment. Aquatic ecosystems are collapsing because they are increasingly denied the water necessary for their survival or they are polluted by human and industrial wastes. International and subnational conflicts are growing because of the inability of our institutions to manage water that crosses political boundaries. And climate change is altering fundamental assumptions about water availability and quality that have guided the past choices we've made about infrastructure, water allocations and rights, and management.

The reality is that we are reaching the limits of the ability of our planet to provide safe, clean water to our burgeoning populations. We can no longer rely solely on the technological solutions we developed and applied in the Second Age of Water to meet our water needs. We must move to a new way of thinking and managing water—what I call the Third Age of Water—by moving to a "soft path for water."

At school, and as I began my professional work, it was always water that drew me. My early graduate work was on the environmental risks of hydroelectric systems and large and small dams. In 1982, I found myself as a representative of the governor of California on an official delegation to China, talking about the environmental consequences of energy development in a country whose air was choking with pollution and whose rivers were already being contaminated or sucked dry for industrial and agricultural uses. I remember visiting massive dam after massive dam proudly displayed by our Chinese hosts and seeing only the dead, dewatered ghosts of rivers, where my hosts saw progress, power, and national development.

My dissertation was one of the earliest efforts to look at the implications of climate change for rivers, streams, soil moisture, and snowpack. After receiving my doctorate, I spent two years on a postdoctoral fellowship looking at the relationships between natural resources and international security and conflict, especially water and conflict. In the early 1990s, I participated in a number of intensive negotiations and discussions about the water resources of the Middle East, especially the Jordan River Basin shared by Syria, Lebanon, Jordan, Israel, and Palestine. On a cold Zurich winter night, I found myself seated between Israelis and Palestinians trying to negotiate the wording of an agreement over water, as a precursor to their 1995 agreement that lays out important principles on water sharing and management.

I visited South Africa as apartheid was pushed out and saw what it meant to the poorest to be denied basic access to safe water. I learned that the failure to meet basic needs for water is not due to a lack of information or technology or money. The failure is both simpler and more complicated: It is a failure of political will, commitment, and governance on the part of the international community, national governments, and local communities. I watched as South Africa's new government rewrote that nation's water laws and Constitution, putting water for people and the environment first. In the late 1990s, I wrote that water is a fundamental legal human right—how could it not be? Without water, there is no life. Without safe water and sanitation, there is no dignity. And the failure to meet basic human needs for water—the failure to provide the minimal amounts of safe water necessary to satisfy needs for drinking, cooking, cleaning, and sanitation—already leads to millions of unnecessary and preventable deaths each year from water-related illness and disease.

These experiences helped show me that while our water challenges can be tackled with technology, technology alone will not solve them. Smart economics will play a role in helping us manage water in a more sustainable way. But economics alone is also not enough. In the end, water is about people and the natural environment. We must look to fundamental changes in thinking, policy, and culture. We must relearn the lessons that indigenous people learned about local waters. We must understand that the solutions to our water problems must look to a more moral and ethical code, and must acknowledge and even embrace the sense of wonder that water stirs in all of us. These are all components of the Third Age of Water. And while fundamental changes and transitions are never easy, I believe they are both necessary and, indeed, inevitable as we move on the path toward a sustainable water future for all. The stories in this book offer insights into this path. They provide glimpses of what a sustainable water future can look like through the eyes of a remarkable set of water warriors, from innovative scientists to fierce defenders of local communities, from poets to pundits, and from all corners of the globe.

THE END OF THE COLORADO

Frank Clifford

Frank Clifford was a staff writer and editor for the *Los Angeles Times* for 25 years, covering government, politics, and the environment. In 2007 he edited a series of stories on the impacts of pollution and overfishing on the world's oceans that won a Pulitzer Prize for explanatory reporting. He is the author of *The Backbone of the World: A Portrait of the Vanishing West Along the Continental Divide.* A native of Minnesota, Clifford lives in Santa Fe, New Mexico, where he works as a freelance writer.

I AM A PROFESSIONAL NAG. I write about the environment. If I give vent to my deepest misgivings I can sound like a sidewalk evangelist warning of Armageddon. In my line of work, as in the preacher's, you can get the feeling that people cross the street to avoid hearing what you have to say.

The Colorado River is certainly a prime topic for dire warnings. If the current drought, now in its tenth year, continues to starve the river, the reservoirs that serve 30 million people in seven states could dry up by the middle of the century. Some scientists say there is a 50 percent chance of that happening.

Can we conserve enough water to avert the crisis and still slake the thirst of Las Vegas, Los Angeles, Phoenix, and Denver? Can we meet legal obligations to provide water to Native Americans and endangered species, while continuing to grow food and generate power? Or is it too late? Are we headed down the road to ruin, to our own version of the Anasazi diaspora?

It is at this point in a reporter's pitch that an editor is likely to interrupt. Weary of doomsday forecasts and hungry for the here and now of the story, he might ask, But what does this waterless ruin look like? Can we go see it? Are there people living in it that we can talk to, today?

Except for the bathtub rings around the reservoirs, there's not much graphic evidence of scarcity, at least not on the American side of the 1,400-mile-long Colorado. Some fields have been fallowed and backyard watering curtailed in many places. But civilization hasn't retreated from the riverbanks.

Yet finding the dust bowl on the Colorado is not an impossible task. You just have to go to Mexico, beyond the border towns, to find it. You head for the salt barrens of the Colorado River Delta, where the spent river emits a thin, bilious heave into the Sea of Cortés. You don't have to conjure up a bleak future there—amid the roiling dust storms, the smell of burning tires, the chemical sloughs, the treeless yards and cheerless huts of half-abandoned villages. The delta brings to mind a saying people have used to describe New Orleans: "the place that care forgot."

Just finding the Colorado down there isn't easy. South of Yuma, Arizona, it looks more like a swamp that has alternately been drained and burned, a maze of sulfurous ponds surrounded by scorched earth—the result of diversionary fires set by smugglers who drive contraband across the shallows in all-terrain vehicles. At the Morelos Dam near Los Algodones, most of what's left of the river is siphoned into irrigation canals and pipelines that carry water 175 miles to Tijuana.

Below the dam, the remaining ribbon of river squeezes between trash-strewn thickets. Signs warn of poisonous snakes. But the river is not dead yet—there are some trickles back into the river that barely salvage it at this point. It is what hydrologists refer to as return flow, or secondhand water laden with effluent from farms, sewers, and treatment plants. It makes up most of the flowage south of the dam. But fish can survive in it, and people still try to make a living catching them.

The fishermen I was interested in are members of a remnant band of Cucapá Indians (spelled Cocopah in the U.S.) whose ancestors came to the delta 1,000 years ago but whose population has receded with the river. Every spring Cucapá fishermen head toward the river's mouth in pursuit of corvina, a popular white fish sold in restaurants and supermarkets. The corvina spawn in the tidal surges that push up the mouth of the Colorado and merge with what's left of the downstream flow. The area, known as the *zona nucleo,* is the core of the lower Colorado River Delta Biosphere Reserve, established in 1993 to protect marine animals, including the corvina, that thrive in the brackish water at the river's mouth.

The Mexican government banned fishing in the core of the reserve to help fish stocks revive. As more and more of the river was dammed or diverted upstream, the environment at the mouth of the Colorado was degraded. By the mid-1980s, scientists reported, two dozen species of fish were headed for extinction and 60 more were at risk.

The Cucapá defied the fishing ban. They believe they have a historic right to fish in the lower Colorado. The argument may be a bit of a stretch. According to José Campoy, the director of the reserve, the Indians traditionally fished much higher in the river and didn't start showing up in the zona nucleo until the early 1990s. Still, the Cucapá have a point. If more water were allowed to flow down the Colorado, and the river supported more fish, the Indians wouldn't have to poach in the core zone.

If the fishermen are spotted by the federal authorities who patrol these waters, they can lose their catch and their boats. For the Cucapá, it is a risk worth taking. A bumper harvest of corvina can reap $5,000 or $6,000, two or three times what many Cucapá families make in a year.

"We cannot afford not to be outlaws," said Andres Lopez, one of the fishermen.

The law isn't the only peril the fishermen face—the spring tides are among the highest in the world. A tidal surge overturned a steamship 100 years ago. Cucapá fishermen have drowned trying to make their way back up the churning river; they did what fishermen are often tempted to do when the corvina are running and they have overloaded their boats.

Although I had visited the Cucapá twice during the past 11 years, I had never gone fishing with them. I would have to do that if I wanted to describe how people live along a dying river.

I had no way of knowing whether the Cucapá would let me go with them except by driving to Mexico and asking them; there was no way to reach them by phone.

With my friend Michael Robinson Chavez, a *Los Angeles Times* photographer, I drove down in mid-March 2008. We followed Mexico's Highway 5 south from the border. It's the same road that ultimately leads to the beaches of San Felipe on the east coast of the Baja Peninsula. Where we were going there were no kiosks selling shrimp and no posh resorts. We brought sleeping bags, packed three ice chests with food in Yuma, and drove about 30 miles south of Mexicali to El Mayor, the tribal headquarters of Mexico's Cucapá Indians.

Spanish explorers and missionaries who traveled up the Colorado River Delta estimated the Cucapá population at around 7,000. Fewer than 200 Cucapá remain. Most of them live in El Mayor, a cluster of cinder block and plywood shacks near the confluence of the Rio Hardy and the Colorado. The residents sell homemade beads, pick cotton, or work in factories around Mexicali. Some sell scrap from wrecked cars they scavenge on the highway. When I was there, about 20 percent of the homes had electricity or running water.

The Cucapá have a tribal historian, a man named Colin Soto, who lives on the U.S. side of the border on a reservation in Somerton, Arizona, just south of Yuma, where most of the Cucapá have moved over the years. They farm, operate a casino, and no longer look to the river for a livelihood. Soto described his kinfolk in Mexico as people without an identity. He said he learned this when he was asked to help a family from El Mayor attend a funeral on the American side of the border. He said the Mexican Cucapá had no documents, no driver's licenses or passports, or any type of information that could be used for temporary IDs.

"We couldn't figure out what to do," Soto said to me. "We told them to bring a bill from the water company. They said, We don't have a water system. We asked them, Don't you have some piece of paper with your address on it? They said, We don't have addresses. We made our own roads and we didn't name them."

We arrived in El Mayor the night before the fishing season began. Working by the light of bonfires, men were repairing nets strung along the dirt streets. Boats and motors were being loaded onto pickups.

El Mayor is within walking distance of the Colorado, but you have to drive south another hour before reaching fishable waters. The Cucapá launch their little fleet of *pangas*—broad-beamed boats about 20 feet long—at a place called San Juan, where they can slide the boats down a dry arroyo into the river. San Juan is not on any map that I have ever seen, and I wouldn't be able to find my way back there. We drove south on Highway 5 and turned off on a nearly invisible dirt track at a signal from one of the fishermen. Then we hurtled across several miles of gray mudflats, careering from one set of tire tracks to another until we jerked to a stop on a high bank overlooking a narrow channel.

Michael and I stood around, waiting for the fishermen to fix a gas line on an old Yamaha outboard motor. They were pirating parts from someone's Chevy pickup. While the men worked, an airplane materialized out of a pastel sky. It was some sort of military aircraft, flying low and slow. The plane banked overhead and made a couple of passes, clearly interested in us.

This place was not like any river delta I'd ever seen, dotted with alluvial islands and tangles of underbrush trembling with bird life. Instead, the San Juan landing presented a scene of mesmerizing desolation—the earth after chemotherapy, bald as a baby's skull cleft by an azure scar, the Colorado River.

The Colorado River Delta didn't begin to look like a desert until relatively recently. When Spanish missionaries first attempted to cross it, they needed the help of native swimmers to ferry them across in baskets.

In the 1920s, when naturalist Aldo Leopold wrote about it in "The Green Lagoons," the delta was still a watery jungle full of jaguars, deer, pronghorn, condors, quail, beavers, and waterfowl. "The river was nowhere and everywhere," Leopold wrote. The delta inspired Leopold's classic comment on wilderness: "Of what avail are 40 freedoms without a blank spot on the map?"

Dams killed the delta of Leopold's day. More than 20 were built, almost all of them in the United States, beginning with Hoover Dam on the Arizona-Nevada border in 1936 and culminating with Glen Canyon Dam in northern Arizona in 1962. After that, almost no fresh water reached the ocean for 20 years. Powerful tides from the Sea of Cortés pushed 30 miles upriver, scouring channels and killing much of the plant and animal life in the delta.

The green lagoons of Leopold's time were the size of Rhode Island. By the early 1980s, these wetlands had been reduced by 90 percent.

American policy toward the Colorado River was rooted in manifest destiny—the notion that Americans had the right to control the continent, as if our dominion were divinely ordained. In 1893, U.S. Attorney General Judson Harmon stated that the United States had no legal obligation to share Colorado River water with Mexico even though the last 75 miles of the river flowed through that country.

A quarter of a century later, when the states that border the Colorado divvied up the river, Mexico was not invited to the table. The U.S. government eventually softened its stand, but only after the Mexican government threatened to cut off water to the Rio Grande Valley in Texas. So, in 1944, the United States agreed to let about 10 percent of the Colorado flow into Mexico. In turn, Mexico directed most of that water to industrial agriculture in the Mexicali Valley. It piped the rest to Tijuana. The needs of the thinly populated delta were ignored. The Cucapá were not consulted.

The Harmon doctrine may seem like a historical footnote today, but its spirit lingers. California, Nevada, and Arizona have set out to capture most of the accidental surplus that flows into Mexico in excess of the 1944 treaty allocation. That extra water, which is often referred to as the slop, has been vital to the survival of the delta's remnant ecosystem. For the past several years, a small network of nonprofit groups on both sides of the border has been attempting to persuade lawmakers in the United States and Mexico to allow some of the slop to flow to the mouth of the river. But the proposal has yet to be adopted.

Three years ago, Congress declared that the United States bore no responsibility for the delta's environment. Officials in Mexico reacted bitterly. "The U.S. has contravened its obligations once again so that it can get more water flowing to its swimming pools and flower gardens," said Alberto Székely, a career ambassador with the Mexican Foreign Service.

But the Mexican government itself is hardly blameless. It hasn't reserved any of the water it receives for the delta, and it wastes a huge portion of its allotment. The government-owned canals and pipelines that deliver water from the Colorado to Mexicali and Tijuana are in such bad shape, officials estimate that more than half of the water is lost or unaccounted for.

Each time I visited El Mayor, a few more houses were empty. There were fewer dogs and chickens foraging in the streets. Yet the village springs back to life in the fishing season. If anything has slowed down the Cucapá migration out of Mexico, it's been the remarkable return of the corvina.

The fish, which is found only in these waters, nearly vanished in the years after the completion of the Glen Canyon Dam when the Colorado stopped flowing to the sea. Then the weather changed. In the 1980s and '90s, El Niño years brought relief from drought, allowing fresh water to reach the delta. The corvina came back, saved by the slop. But its future was not assured. The lack of fresh water in the delta combined with fishing pressure has led conservationists to urge consumers to avoid corvina.

The corvina is a member of the croaker family, and if you have ever gone fishing for corvina, you understand how the family name originated. The noise that comes from a boatload of dying corvina can only be described as a lamentation, a full-throated chorus of croaking.

The Cucapá set out from the San Juan landing in a flotilla of pangas. No one wears a life jacket. If the motors stall, which is not unusual, the fishermen can only hope they are close enough to shore to escape the pull of the outbound tide. If they are swept away, the Cucapá, who don't carry emergency communication equipment, can face hours or even days on the open ocean.

The day I went with them started slowly. The morning's catch was disappointing. We hauled in fewer than 50 fish. That afternoon we sped even farther south, stopping off Montague Island, near the mouth, in the heart of the forbidden zone. That's where the fish were. I was in a boat manned by Julio Figueroa and two teenage helpers.

The Cucapá use gill nets to trap the corvina, then pull them in hand over hand, detach the fish, and throw them into the bottom of the boat. When the catch is modest, there is time to free each fish from the net. But when you are in the middle of a run, and hundreds of fish are getting snared, it's all you can do to retrieve the heavy nets. Within an hour we were standing on a ton of fish. First the bottom of the boat was covered, then the seats. It was hard to hear anyone speak over the din of the corvina death rattle.

When I suggested that the boat was getting a little too full and that maybe we ought to call it a day before we foundered, I got a couple of disdainful looks and no reply. The old gringo was losing his nerve, they figured, and I was. They kept pulling in more and more fish until Julio abruptly waved his hand and signaled them to stop. "Who has a knife?" he asked. He wanted to cut the net before we were entombed in corvina. There was still 25 yards of net in the water, with enough fish in it to prevent us from moving or to capsize us if we tried. No one had a knife, but after a frantic search, one of Julio's crew found a metal file. After ten minutes of sawing, we were free and clear.

Julio pulled on the start rope. The motor coughed and died. He adjusted the gas flow. Still no luck. We drifted. Around one more bend of the island and the ocean would be in sight. The tide was going out, and the afternoon breeze had turned into a steady blow. After another five minutes of yanking on the rope, the motor fired and we started north, but very slowly. We were bucking whitecaps. Pangas are stout, seaworthy boats. Their spreading bows are designed to disperse the spray from oncoming waves. But we were riding too low in the water. Instead of rising on the waves, the bow was plowing through them and water was cascading over the top. Julio told us to move as many of the fish as we could toward the stern. It was like using your hands to move a small mountain of wet cement. Julio motioned for all of us to move back and get as close to him in the stern as we could. He himself was standing in ten inches of water.

By now the tide was beginning to recede. The river was a foot lower than it had been in the morning. Besides capsizing or being arrested, the worst thing that can happen to a Cucapá fisherman is to be mired in a dry river bottom several miles from the landing, his fish rotting in the afternoon sun. We rounded a bend and saw that the boat Michael was in had become lodged on a sandbar. The only way to get off it was for the crew to jettison a substantial portion of their catch. That's what they were starting to do as we passed them.

Julio's two helpers offered a hand. He and I kept heading upriver, Julio muttering to himself *"malo viento"* ("bad wind"). Against this evil wind the heavily laden boat was barely able to make forward progress. Julio tapped me on the shoulder. "Look there," he said, pointing to our left. A smaller boat that had been running parallel with us was in trouble. Its bow had submerged, and as we watched, the front half slipped into the river. Its three occupants were huddled in the stern. The panga rolled to one side, and then the men were in the water, clinging to the sides of the boat. Julio's stepson was one of them. But Julio said there was nothing we could do. If we tried to turn back, we would slip sideways in the waves and tip over. So, we pushed on upriver.

As the afternoon faded, the headwind slowly died and the river calmed down, allowing us to pick up speed. But we could not travel fast enough to beat the outbound tide. About a mile short of the landing, the river ran out of water. At that point, we had two choices. Wait on board for six hours before a rising tide would bring enough water to float the boat, or we could walk ashore.

Julio opted to stay on the boat. The sun was setting, and he wouldn't have to worry about the fish rotting. But I didn't relish the idea of sitting in wet clothes for six cold hours. Getting to shore would require wading through a quarter mile of muck while holding onto a rope that was tied to a pickup truck parked on a nearby bluff. That didn't look so hard. I watched as a young boy came out with the rope. He was half crawling but seemed to move easily enough.

I stepped out of the boat and immediately sank up to my waist. As I struggled to free one leg and then the other, I lost both shoes. If I were going to move at all, I was going to have to travel on my belly, grasping the rope and wriggling forward like some primeval amphibian.

I tried to make myself weightless, but the "quicksandy" river bottom didn't just suck, it scraped and clung. It was the salt. Less than halfway to the waiting truck, I realized the front of my windbreaker had been shredded. I lay my face on the cool mud and took a breather. This is the way life on terra firma began, I imagined. This is how we emerged from the ooze.

Surely, I thought, the Cucapá could find an easier way to make a living. Even some of their own kin, across the border in Arizona, thought they were crazy to go on like this. Didn't they want a better life? Didn't they understand that here they were doomed?

Right now, Julio Figueroa could be sleeping in a warm bed in Somerton, instead of sitting on a pile of dead fish in a dry riverbed waiting for the morning tide to rescue him.

I looked back at Julio's hunched silhouette, a profile of courage and obduracy. But he is not so different. Society may encourage adaptability, but it reveres tenacity. Up and down the Colorado River, in the United States as well as Mexico, people are fighting for their share of a bounty that the river probably cannot continue to provide. For some farmers and ranchers in California's Imperial Valley and Nevada's Great Basin, losing the fight will spell the end of a way of life, as their water is piped to San Diego or Las Vegas. But they won't go quietly.

Because of our environmental laws, nature may fare better along the U.S. side of the Colorado than it has in the delta, although proposals to pipe water to cities from rural areas of Colorado, New Mexico, Utah, Nevada, and California could turn lush basins into lifeless deserts.

I resumed my crawl through the mud. Slowly, the ground began to harden. My feet found a purchase. I stood up, sockless now, my jeans in tatters.

I heard Andres Lopez laughing. "I don't believe it," he said. "The old gringo made it."

He threw his jacket over my shoulders and led me to a bonfire, where I found Michael safely returned with news that the crew of the capsized panga had also made it to shore. We passed around a bottle of warm beer and toasted the day's luck. Then the others began loading boats and fish onto the waiting trucks. They planned to sell their catch later that day to buyers from Mexicali and San Luis.

For Michael and me the fishing season was over. But for the Cucapá, it was just beginning. They planned to return every day that the corvina were still running. They often fish well into May, when the juvenile corvina tend to show up in large numbers. That's when fishing can do the most damage to a species by pulling out the youngest before they are old enough to reproduce.

When I asked Andres about that, he shrugged, as if to say, You do what you gotta do.

Later that spring, the federal police stepped in, impounding several tons of fish. I haven't talked to Andres or any of the other fishermen about what happened, but I suspect they'd say that their rights were violated, that indigenous people are entitled to make a living the same way their ancestors did, and that the Cucapá's little fleet of pangas, about 30 boats, won't drive the corvina to extinction.

The courts in Mexico have repeatedly ruled against the Cucapá, saying that they are not exempt from the ban on fishing in the core zone, according to José Campoy, the director of the biosphere reserve. But the Cucapá's arguments are not entirely without merit. Whatever impact they do have on the delta, it is small compared with the cumulative effects of the upstream dams and irrigation projects that have starved the delta of fresh water and pit a vanishing tribe of Indians against an imperiled species of fish.

Campoy said efforts are being made to find the Cucapá other ways of supporting themselves. But he conceded that weaning them off the spring run of corvina won't be easy, not as long as there is a market for the fish.

"As long as they are poor and need the money, some of them are going to be out there."

On our way back to the United Sates, Michael and I stopped at a Mexicali restaurant for lunch. It was a bit of a celebratory meal. We had gone fishing with the Cucapá and lived to tell about it.

I ordered seafood and was nearly finished before I realized what I was eating. It was very good.

A RIVER STORY

Marion Stoddart

Marion Stoddart was born May 26, 1928, in Reno, Nevada, the daughter of Idaho homesteaders Atlee and Ruth Jackson. In 1949 she received a B.A. from Occidental College majoring in sociology and anthropology. She began her devotion to conservation and the environment while teaching at a pioneer School Conservation Camp in California. In recognition of her lifelong work, Stoddart has received many awards including the United Nations Global 500 Award in Nairobi, Kenya (1987), the U.S. Environmental Protection Agency Environmental Award (1972), and a presidential commendation (1970), and was an honoree of the National Women's History Project as one of the "Women Taking the Lead to Save Our Planet" (2009). She was featured in *National Geographic*'s 1993 Water Issue and Lynne Cherry's 1992 award-winning children's book *A River Ran Wild.* Stoddart lives with her husband, Hugh Stoddart, in Groton, Massachusetts. They have three children and five grandchildren.

WHAT IF I TOLD YOU my keenest dream as a young girl in the wild Nevada desert was to root myself in a small colonial Massachusetts suburb. Would you believe me?

Well, neither would I. But that is where I found myself in 1962, the year we moved to Groton, Massachusetts, to be near my husband's work. I was a mother of three and the wife of a gifted scientist and inventor, and I was there for all the right reasons: love, loyalty, family. I was just 34 when, shortly after our move to Groton, my husband told me something I never expected. He said the town was perfect and he'd be happy to live there for the rest of his life. Didn't I agree?

I did not. To the very bedrock of my being, I did not!

I have always loved new people, new places, and new adventures. In Fernley, the little desert town in Nevada where I grew up, I used to wait for the Greyhound bus to stop at my father's general store. As people stepped out of the bus, I'd study the different faces, fascinated by the world of possibility they implied. On Sundays, I'd escape the indoors and ride my Mustang pony through the desert sage for miles and miles, always on the lookout for darting jackrabbits or diamondback rattlers, dreaming of the countless adventures I could embark upon. As a girl and young woman, when my spirits felt caged, imagining a world of possibilities provided me with a profound sense of freedom and independence.

I never thought for a second I'd spend the rest of my life in one place, in one community of people, let alone a quiet town in Massachusetts.

Here in Groton the land was so different—but my instinct was the same. I sought in nature what I could never find indoors—balance, belonging, a path. I desperately wanted to find meaning. How could I be of use? How could I reconcile myself to living in this little country town forever? I knew my husband and children needed me, but what else could I offer? What was the point of *me* in the world?

The answer came to me on the wind. I'm not being poetic. I am talking about a foul stench so powerful it could peel paint. It wafted behind our house. It permeated the town—even the gates of the prestigious Groton School, alma mater of President Franklin Roosevelt, could not keep it out. The stench was the smell of the Nashua River.

Or rather, what *once* had been a river. The Nashua was once a magnificent 56-mile tributary—home of the Algonquin Indians, who called it *Nash-a-way*, or "river with the beautiful pebbled bottom." Now the bottom was obscured by deep, stinking layers of immovable sludge. The water once teemed with a variety of fish and waterfowl. Now it was a putrid repository for industrial waste, raw human sewage, and unnaturally brilliant dyes from the paper factories upstream.

It was perfectly legal to pollute the river in those days. Not only was there no federal agency guarding the purity of natural resources, but what local laws did exist actually favored the rights of factories and cities to dump their refuse! Clearly, the river had no ally. So I decided *I* would be its ally. I would devote my life to restoring the Nashua.

I had no idea how I was going to do it, nor any a clue how long it would take. But it didn't matter. I'd been looking for a challenge that would fill my lifetime—here was the perfect candidate, literally in my own backyard. The moment I decided, I felt a sense of calm. Groton began to feel a little more like home.

Most people want their lives to make a difference. We all seek purpose and meaningful work. But the problems in our world can seem too daunting to one individual. It's easy to give up before you've even started.

One day, after I'd decided that restoring the Nashua would be my life's work, I heard a radio program in which the speaker said, "One person can do the work of 1,000." I felt an almost electric thrill. The concept galvanized me. I was only one person, to be sure—but I'd just been told I could do the work of 1,000. Suddenly it didn't matter that I didn't have all the answers. What mattered was that this disembodied voice from the radio had found me, bringing with it just the rallying cry I needed.

Ever since my desert girlhood, I'd had a keen awareness of water. Our family alfalfa farm was on the first reclaimed land in this country, what once had been an inland sea—full of nutrients, lacking only water. Like all farmers in the area, we paid dearly—by the acre-foot—for every drop of water that came to us via the Truckee-Carson Irrigation District, all the way from Lake Tahoe.

But it wasn't just water's effect on our family's livelihood that I noticed. You can't admire the tenacity of desert creatures, like my favorite horned toad that shoots blood from its eyes to frighten predators away, without also opening your heart to their often poignant and perennial struggles for water. In this setting, it is easy to understand how bodies of water—rivers, streams, lakes, and oceans—have been revered by native cultures as the greatest givers of life, the mothers to us all.

For me, nature was always my closest companion. Now that I was a mother myself, I felt protective of the devastated river behind my house the way I felt protective of my children. It deserved my love and respect and all the passion and ingenuity I could muster. I yearned to see it thrive and flourish just as I yearned to see my children crawl and walk and run.

So I picked up where I'd left off.

Years before, as a new mother in another small Massachusetts town where we'd lived briefly before settling in Groton, I'd become involved in water and land conservation through the League of Women Voters. The League of Women Voters was my way of staying in touch with the adult world. We young mothers—fresh from college, barely more than newlyweds, unprepared for the loneliness and isolation of our new existence off the intellectual grid—supported each other. One of us would take a turn babysitting all five or seven toddlers while the others went to League meetings, then we'd rotate. It preserved our sanity. And it taught me the politics of conservation: the processes for getting legislation written, sponsored, and passed. The Sudbury League was instrumental in promoting a state bill that protected hundreds of acres along three area rivers. It was a heady success and an essential education.

My first thought was to approach the Nashua the same way we had in the Sudbury League of Women Voters: acquire land along its banks and protect it forever in conservation. I started with a map and marked all the towns along the Nashua, intent on learning who the conservation and political leaders were in each town. Then I went to Boston to meet with heads of state and federal conservation agencies and urge them to buy land along the Nashua.

But there was a catch-22. The state informed me there was no policy in place that enabled Massachusetts to acquire land along polluted rivers. Reluctantly, I put aside my vision of a lush and thriving greenway the full length of the Nashua River for now. Another battle loomed, and I knew the Nashua wouldn't be safe until I won it.

It was 1966. A conservation commissioner in neighboring New Hampshire wrote a letter protesting the "nuisance condition" of the Nashua—language that sugarcoated the river's designation as one of the ten most polluted rivers in America—and sent it to his governor alongside a petition with 600 names. In politics, numbers talk. When legislators saw that many constituents converging on one issue, they leaped to the fray like spawning salmon.

In Massachusetts, I was motivated to follow our neighbor's example. Despite my preference to save the land first, it was clear the river cleanup movement had more momentum. In 1969, Ohio's Cuyahoga would famously burn, setting off fires of protest all over the country. But even three years earlier, the outrage of ordinary people over the desecration of their rivers and other natural waterways had passed a threshold. I took advantage of these isolated pockets of outrage to build an alliance, the Nashua River Clean-Up Committee, working toward one goal. After educating myself on the water pollution issues, I was able to educate community leaders, stakeholders, and ordinary citizens. Gradually, we had a diverse group of thousands who all believed that change was possible.

The momentum we gathered was evident when we went to Beacon Hill to petition our governor to clean the Nashua. We brought with us a petition with more than 6,000 names and a jar of river sludge. Mayors and selectmen from every town along the river took part in the presentation, which impressed Governor John Volpe perhaps more than anything else that day. He couldn't believe so many important officials had taken time away from their busy jobs—and he vowed to keep that mason jar on his desk as a reminder of what had to be done.

We had the governor's, and the state legislators', attention. Now we had to get the attention of Senator Edward Kennedy. He had arranged to bring Stewart Udall, Secretary of the Interior, to tour New England's polluted rivers, and we persuaded him to include the Nashua. There was a new rating system for rivers: The highest grade was A, fit for drinking. B meant the river was suitable for water sports like canoeing, fishing, and swimming. The lowest grade was D, suitable for transporting waste. The Nashua's grade? U—unfit even to transport waste.

With two days' notice, we rallied 500 people to greet the entourage at a local airport. I sat on the platform with pad and pencil, struggling to decide how much I wanted to ask for on behalf of the Nashua. Lieutenant Governor Elliot Richardson, who would gain international fame just a few years later as attorney general during the Watergate hearings, peered over my shoulder and saw me struggling. He knew I'd be addressing the crowd, Senator Kennedy, and Secretary Udall in just a few minutes. No sooner did I scrawl a wish list for the river—including swimming—than I erased swimming in a flurry of pink eraser crumbs. I was afraid to ask for too much! As of now, the river was an open sewer filled with untreated excrement, chemicals, and paper waste. Every day it was a new color, depending on the whims of the paper mills:

yellow, green, gray, white, and red. In the midst of a three-year drought, it didn't even flow—birds and small animals could literally walk across it without getting their feet wet. What was I thinking, imagining children splashing in the river. I didn't want to lose my credibility by seeming naive or unrealistic.

I felt the lieutenant governor looking over my shoulder. "Ask for swimming," he whispered. "If you don't, you'll never get it."

It was one of the most important messages I've ever received. Absolutely liberating. Since then, I have always asked for what I want, not for what I was willing to settle. Later that same day, when Governor Volpe returned to Boston, he signed into law the Massachusetts Clean Waters Act—a crucial first step. Massachusetts was the first state in the nation to pass such a bill.

The next step was on the national level. Once the Federal Water Pollution Control Act became law, area towns and businesses were obligated to build wastewater treatment plants to clean up the Nashua. But how clean was clean? Factions emerged and argued over this point. The Department of the Interior had issued a deadline: All states must submit their standards for interstate rivers after conducting public hearings. This was the first time in the history of our country that citizens were given the legal opportunity to say how clean we wanted our interstate rivers to be. The Nashua River Clean-Up Committee worked for a year to identify all of the organizations in the watershed and asked them to prepare a statement for the hearing. Our hearing was scheduled at a nearby state college. Predictably, local industries sided with the city of Fitchburg seeking the least costly cleanup for a grade D classification—good enough to transport waste. State officials said they'd settle for water clean enough for irrigation but not water contact sports: a grade C. Citizens, however, asked for a B: swimming, boating, fishing, irrigation.

We were joined by what at first seemed an unlikely ally: the local unions. They were furious with one of the nearby industries for threatening union members' jobs if they favored clean water. We filed a citizen suit to force this company to honor its obligations. We won our grade B classification for the Nashua because hundreds of ordinary people were willing to stand up and be counted at that public hearing. And we'd had the great good fortune of finding an unexpected friend; by threatening a lawsuit, we were assured that one of the Nashua's biggest polluters would remain in its contractual agreement with the city of Fitchburg to help construct an industrial wastewater treatment plant.

Over and over again, I have found help where it's least expected.

Just as friends turn up in unlikely places, so do naysayers. But negative thinkers can drain the soul out of any endeavor. I learned long ago only to associate with positive-thinking people—the energy is contagious. You will never change the mind of a negative person, any more than you can change the mind of a rattlesnake. Learn to hear the rattle, step aside, and move on.

Just as I learned to sidestep negative people, I learned how to deal with hearing no. If I had listened every time I heard the word "no," I would have given up on the Nashua before I had started. Throughout my work with the Nashua, a negative reaction was always an exciting challenge for me. It simply meant finding another person who *would* say yes, or another route to solving a problem.

In the years after the passage of the 1965 Federal Water Pollution Control Act and the 1966 Massachusetts Clean Waters Act, environmental reform came fast and furious. The Environmental Protection Agency was formed in 1970. The Federal Clean Water Act was passed in 1972. And by that year, after millions of dollars in state and federal funds were spent on the construction of wastewater treatment facilities, the cleanup of the Nashua was essentially done.

Now what was I to do?

The volunteer Nashua River Clean-Up Committee I'd started in the mid-1960s had evolved into the Nashua River Watershed Association, of which I became the first executive director. Finally, we could gather the forces of all the conservation organizations in the watershed along with help from state and federal agencies to acquire land or rights to land for greenway that I'd always envisioned. This greenway, now 60 percent complete, protects water quality, wildlife habitat and corridors, recreational trails, and access to the river for the benefit of all. It was vital to me that the association be sustainable as an organization, and not overly dependent on one person, so in 1977 we found an executive director to encourage development of new leadership.

The river had been restored much sooner than the lifetime of work I had imagined. It was time for another life reassessment!

During the past decade of intense advocacy I'd always relied on my partner on the home front, my husband, Hugh. He's the one who collected a jar of river sludge to take to the State House. He kept the hungry mouths fed at home with countless tuna casseroles. He was our committee's bookkeeper and—conveniently—biggest financial donor. But he'd had a mixed relationship with the river I'd chosen. For years, every time he rode his bike over the river he would stop on the bridge and throw in a rock.

"Marion has a lover," he'd say. "It's the Nashua River."

Hugh knew from the start that I would never be an indoor girl—he'd never asked me to be. But it was time for me to contribute to the family coffers. Our children were entering college. The money I'd earned giving tennis lessons during the early days of the river cleanup—never a king's ransom, to be sure—had gone directly to stamps for our mailings. How could I support the ongoing land purchases of the watershed association, help Hugh support our family, and meet new and diverse people—all while spending as many waking hours as possible out of doors?

In the end, I started a business—Outdoor Vacations for Women over 40. Over the past 25 years I have traveled with more than 3,000 women on hundreds of trips all over the world. I continue to volunteer for the Nashua River Watershed Association, and have been enormously gratified to contribute to its financial viability.

Forty-seven years after a crisis of the spirit led me to a simple resolve—"I want to restore the Nashua" — the river keeps running. Today canoeists, and picnickers, and schoolchildren who visit with their science classes love it. It is a delight to swim in, cold and bracing, mysterious and thrilling with the quicksilver bodies of native fish darting underneath. And when Hugh and I bicycle across the Nashua together, he no longer throws rocks into its thriving depths.

My vision for the future is that people around the world will take ownership of the rivers in their communities. That they will learn to love them, respect them, and protect them. Every river, no matter how polluted, has the potential to be teeming with fish and waterfowl, bordered by natural habitat and corridors for wildlife, life sustaining in every way.

When I moved to Groton, Massachusetts, I felt that I was isolated and insignificant, that a whole world of excitement and opportunities existed beyond the perfectly mowed lawns of my small-town life. In retrospect, I have to agree with my husband, Hugh, that this community *was* the perfect place for us to live for the rest of our lives. In devoting myself to preserving the Nashua River, I rediscovered a stronger sense of meaning in my life. I hope that my story of the Nashua River will inspire thousands more to engage in change that they believe in, however insurmountable the odds may seem.

Finally, and most important, I cannot thank enough all the dedicated supporters—the real ones who made the journey possible.

LIQUID ASSETS ON STEEP SLOPES

Anupam Mishra

Born in 1947, Anupam Mishra spent his childhood in Gandhian communities across central India. He moved to Delhi with his family after All India Radio employed his father, a renowned poet, and in 1969 obtained a master's degree in Sanskrit from the University of Delhi. A Gandhian and environmental activist, Mishra has spent decades in the field of environmental protection and water conservation and is among the most knowledgeable people about traditional water-harvesting systems in India. Mishra has been associated with the Gandhi Peace Foundation since its inception and is the winner of the Indira Gandhi National Environment Award. He has written two books on traditional water management and water-harvesting systems in India, *Aaj Bhi Khare Hain Talaab (Ponds Are Still Relevant)* and *Rajasthan Ki Rajat Boonde (The Radiant Raindrops of Rajasthan)*. Mishra lives in Delhi with his wife and son and edits the periodical *Gandhi Marg* for the Gandhi Peace Foundation.

"THE WATER IN SPRINGS of my hills is cool. Do not migrate from this land, o my beloved."

In the central square of a village called Daund, deep in the heart of the Indian Himalaya, a group of 15 young girls danced to this lyric. The villagers sat watching, undeterred by the heavy June showers of the monsoon that had just arrived. Among those gathered were many young and old women, their fathers, brothers, husbands, and sons having spent several springs in *pardes,* which literally means "foreign" but in this case means "the plains of northern India." There were also old men, retired to the hills after spending years working on the plains. On drenched woven cotton rugs sat young children—those most likely to migrate out of the hills. Did the lyrics of the dancers' songs have the pull to stop mass migration from this Himalayan village to cities like Delhi?

This village is small enough to not make a mark on any map of the Himalaya. Located 6,000 feet above sea level, it is removed from the rest of the country. Even the stream flowing deep in the valley seems little more than a thin line that gets obscured by sheets of rain and blankets of fog.

If your curiosity compels you to seek out this village, you first need to journey to that part of the western Himalaya that was carved out of the country's most populous state, Uttar Pradesh, and made into the new province of Uttarakhand. When you reach Jim Corbett National Park, along the Ramganga River, travel upstream and you will hit the Doodhatoli Range, rising up to 11,000 feet. *Doodhatoli* means "land of milk," a name that characterizes the ecology here in these pastures above the tree line. Nestled in this range is Daund village. Here the clouds recede, but the water does not. The upper reaches send it down with velocity. Every drop erodes a little bit of the soil and carries it into the stream that will join the Ramganga and make the soil into the silt of Corbett National Park.

The dance troupe performing in Daund packs up its instruments and prepares for the next stop on its road show. Today it's Daund, tomorrow the village of Dulmot, then Janadriya or Ufrainkhal. This is no vaudeville troupe. It does feature a few dancers, singers, musicians, and country-made musical instruments. What the show does not feature is the hundreds of implements like spades and picks that labor harder than the troupe to slow down the water gushing downhill, to hold together the soil—all this to revive the forest and farming that has suffered years of neglect. The instruments—and the implements—are attempting to bring back the melody and the rhythm of ecology to the cacophony of mindless development that has overwhelmed the Himalaya.

This alignment of culture and ecology started in the village of Ufrainkhal, in the Pauri Garhwal district of the province of Uttarakhand. That was 25 years ago. Today, it spans 136 villages. Its aim: creating an atmosphere of conserving ecology, to get people to rediscover that their lives cannot improve without such an improvement. That includes getting them to tend to their forests, their water sources, their pastures, their fuel sources, and their dignity.

Sachidanand Bharati is the leader of this troupe. He teaches at the local college. His own education, however, was in the neighboring district, best known for its Chipko (Hug the Trees) movement: In the early 1970s, some of the neighboring villagers had confronted contractors with permits to log their surrounding forests. The villagers knew well that hill slopes do not obey government land records. If the forest department's land was deforested, they would face landslides, flash floods, and, eventually, water scarcity. To save their home, villagers protested the logging by wrapping their arms around surrounding trees, literally hugging them. The contractors lost their nerve in the face of entire villages showing the kind of nonviolent commitment that Mohandas Gandhi's troops showed during the struggle for independence from British rule. Soon after, the Chipko movement became a symbol of popular environmental conservation in the face of the state's ecological shortsightedness.

During his college years, Bharati got a crash course in environmental management as an associate of Chipko leader Chandi Prasad Bhatt. He experienced firsthand how a popular nonviolent movement could both stop deforestation—the government was impelled to scrap the logging leases and declare a decade-long moratorium on logging—and inspire people to plant more trees, to regenerate their forest. His efforts resulted in a student group, whose name, translated into English, means "Friends of the Trees."

Bharati graduated from college in 1979 with a realization that protest and constructive efforts go hand in hand. He returned home to find that the state's forest department had declared a logging moratorium in the village as a result of the Chipko movement but had granted fresh logging leases in the forests around the neighboring village. The contractors were eyeing fir trees (*Abies concolor,* a slow-growing species that supports a diversity of life in its undergrowth).

Bharati's training, tact, and temperament were suited to this challenge. He got together some friends and went from village to village, talking to people in a calm voice that persuaded but did not agitate. Villagers could see the sense in the simple message the local boy delivered, which was essential for those living in the Himalayan ecosystem: Although the forest might stand on government land, its felling would bring destruction to their doorstep. His message continued: If we stand together, the forest will remain standing. His tone and delivery—as well as the truth of his words—resonated with the villagers.

But dealing with the government officials—known for their arrogance and corruption—required tact. Bharati's calm approach, backed by the strength of the support he had mobilized, persuaded a senior official to send up a team to see if the terrain was suitable for logging. The government faced what the villagers had encountered shortly before: a man armed with truth. The inquiry team agreed with Bharati's claim. The logging leases were scrapped.

The villagers learned two lessons from Bharati even before he took up a teacher's job: One, a united village could resist bureaucratic power and reverse unfavorable government decisions. Two, if the villagers could prevent ecological destruction, they could also join forces to regenerate their forests. Bharati decided to hold a two-day environment camp, inviting neighboring villagers.

There was no road going to the village, then (there is an unpaved one now). No means of communication, no funds to gather the people, spread far and wide, across difficult terrain. Additionally, those who came would have to be fed and lodged. Bharati wrote a letter to New Delhi's Gandhi Peace Foundation, which had been the first to report on and support the Chipko movement. The response was quick: a money order for 1,000 rupees (at that time about $70).

July 1980 saw the first environment camp in the Doodhatoli mountains. Villagers reported on the state of the surrounding forests, exchanging notes on legal and illegal logging that had carried on silently. The state of Doodhatoli's forests was no longer secret. The camp ended with the planting of seedlings and saplings. The camp had also planted an idea, although the hands planting the saplings did not know the idea would become a large tree under which many other constructive ideas would germinate.

Doodhatoli Lok Vikas Sansthan was formed in March 1982, a small organization with no budget. Bharati's approach was written into the group's charter. It would not ask for government or foreign funds but would rely on the resources of the people whose survival depended on the hill ecology. It would take the organization another 13 years to take up water conservation on a larger scale. In the beginning, it was primarily about forests; the forest department nurseries offered saplings of commercially viable trees like pine, which are of no use to the hill ecology or to the village economy. This meant that Bharati's group had to create its own nurseries, which required people to collect seeds. Children and women were recruited for the job—for no payment and no benefits. Volunteers knew the dividend would accrue sometime in the future.

Nurseries require water, which was becoming scarce, especially in the summer months when seeds germinate. Summer in these hills is the season of forest fires, primarily because of the pine trees the forest department had planted for its sap, which it harvested for turpentine. Pine needles stack up on the floor and wait for the slightest spark, which sets hill after hill ablaze. The fires here also consume the natural forests, which nurture more diversity than pine plantations and hold healthy levels of water in the soil. The villagers reeled under a vicious cycle: Lack of soil moisture made the forest vulnerable to fires, and fires smoked out the trees that could hold moisture in the soil. Breaking this cycle required an engineering intervention.

The local boy decided to look locally. He had read about age-old water conservation systems in the Himalaya, which varied according to the slopes. Cultured over the centuries, these were the work of people who had observed the interplay of water, soil, vegetation, and gravity. The answer lay closer than they had imagined, in the village's name: Ufrainkhal. While *Ufrain* is the name of a goddess, the suffix *khal* refers to a type of pool characteristic of this region. It is smaller than a *taal* (lake) but bigger than a *chaal,* which is a series of very small pools along a slope. Several villages and towns in this region carried such suffixes, showing that habitation was built around water conservation—a village in the neighboring Tehri Garhwal district is called Sahasratal, which means "1,000 lakes."

The villagers, though, had forgotten the relevance of this nomenclature, the relevance of pools in the names of their habitat, and the pools' relevance to their survival. For this they paid a heavy price in land and forest degradation. Floods and drought had become a part of the annual cycle, soil erosion an everyday affair. When villagers had even forgotten the meaning of their village name, there was no hope of finding the method of making these bodies of water. With no examples to follow, Bharati decided to experiment. The people who had devised the form of these pools were his own ancestors.

Bharati began with the smallest form: the chaal. It was suitable for the steep slopes of Ufrainkhal, as its small size allowed water to be retained in small quantities, without succumbing to gravity's demands. The Doodhatoli group experimented with varying shapes and sizes in the early 1990s. People accustomed to soil and water management in their fields did not take long to settle on a calibrated proportion for the chains of pools they had in mind. From 1993 to 1998, the pools they'd envisioned became reality on the slopes.

The first dramatic impact was on a small river that had once flowed down to the valley. Years and years ago—nobody can remember when—the name of this river was changed to Sukharaula, meaning "dry channel." In 1994, water once again appeared in the riverbed and ran for a few months after the rainy season. Each subsequent year saw greater and longer water retention in the river. By 2001, it had acquired the shape of a full-fledged seasonal river. The villagers felt it was time to rename the river. They called it Gadganga, combining the name of the village on its bank, Gadkharak, with that of the holy Ganga. This rivulet is a tributary of the river Pasol. Its newfound robustness added to the Pasol's flow.

While the villagers invested effort in creating the pools that retained so much water, nature responded with its invisible efforts. The vegetation began changing around the villages where chaals were dug—in the forests and in the fields. The vegetation multiplied the water-retaining effect of chaals.

In 2000-01, the newly created state of Uttarakhand faced a severe drought, which exacerbated the annual phenomenon of forest fires. Up to 80,000 hectares of forests burned in the state that year. But the villages in and around Ufrainkhal did not burn due to their new water-pooling practices. Fortified with the additional moisture in the soil, the healthy vegetation offered stiff resistance to fire. So did the villagers. Yet three women who worked with the Doodhatoli Lok Vikas Sansthan died fighting fires in government forests. They had taken water from their chaals to put out the fires, because they feared the inferno would soon reach their lands and forests. The hundreds of villagers who fought the fires here—like the three women who paid the ultimate price—did not have the benefit of a privileged education, but they had learned an ecological lesson that consistently eludes highly educated people in the parts of the world considered far more developed (think of forest fires in California, Greece, and Australia).

The villagers' efforts benefited a government scheme, too. The state government had installed pipes to supply drinking water to villages from hilltop springs. Although some water sources dried up, the installations around Ufrainkhal consistently found water to pipe.

A few years earlier, the government had built an office building to start a watershed development plan above Ufrainkhal. Bharati wrote a letter to the authorities, saying the village did not need the government's largesse, as it was able to satisfy its own needs. A government team visited the village and affirmed this claim, and the watershed plan was withdrawn. The building instead was used to provide a shed for cattle and goats. The forest, in the meantime, had begun to do better.

Bharati's troops have built 12,000 chaals in 136 villages to date. Within these areas, there are several patches of thick forests, varying in size from 30 hectares to 300 hectares. In several parts of these forests the areas the villagers have regenerated are healthier even than the government's specially preserved forests—those of the villagers have a greater diversity of vegetation in them, with several broad-leaved trees like oak, alder, rhododendron, and fir. The canopy is usually 100 feet high, in general. The ground covering is several inches thick, with a springy texture that makes walking difficult. It is safer to walk the trodden path in these forests for another reason: Wild animals thrive in forests regenerated by this rural waterworks compared with the protected forests of the government.

The cadre that has brought about this transformation is well worth an introduction, because Doodhatoli Lok Vikas Sansthan does not have a regular budget, does not have any funding from any government or nongovernmental entity, and is not supported by any nonprofit. Some well-wishers send in a check once in a while. Annual expenditure seldom exceeds 25,000 rupees.

The organization does not have any full-time staff, though it works full-time; three associates of Bharati's form the core. There is Devi Dayal, a postman who has to walk through villages to deliver mail, for there are no automobiles or passable roads here. Along his route, he observes the forests, gathers information, and delivers ecological messages (without charging postage). There is Dinesh, a medical practitioner trained in the Ayurvedic system of traditional medicine. Like Devi Dayal's, his line of work involves meeting many people and talking to them. He wraps medical remedies in messages aimed at healing social and ecological relationships. The quartet is completed by Vikram Singh, who runs a small grocery in the neighboring village. His merchandise comes packaged with social provisions, and his shop is a hub of conversation and social exchange in a region where large community halls are impossible to build.

This quartet maintains a regular communication with about two dozen volunteers in each village. Invariably, they are women, for the men migrate to the plains for employment. In the work of Doodhatoli Lok Vikas Sansthan, they see hope of a prosperity that would allow their husbands, sons, and brothers to stay in the village, much as the chaals retain the water. The thousands of chaals built here and the hundreds of hectares of regenerated forests are their only hope. They guard these waterworks and the forests like mothers guard their broods. They number in the hundreds, though their names are not on any roster.

They have a simple way of handing over forest protection duties to the next shift—typical of how the women here combine music and rhythm in daily chores. The woman in charge of forest protection for the day carries a baton with a string of mini-bells tied on top. The sound of the bells works like Morse code across the hill forests. When a woman is done with her shift, she returns to the village and leaves the baton at the doorstep of a neighbor. Whoever sees the baton in front of her house takes up the guard duties the following day, no questions asked.

This is the routine. It's broken by the periodic environmental camps, which all the women turn up for. There is song and dance, the same song and dance now made stronger by the ecological notes:

"The water in springs of my hills is cool. Do not migrate from this land, o my beloved."

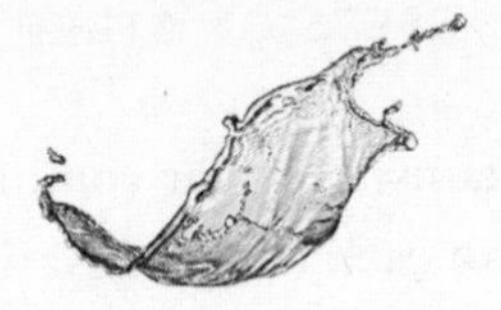

HONEST HOPE

Sandra Postel

Sandra Postel directs the independent Global Water Policy Project, which aims to promote the preservation and sustainable use of Earth's fresh water through research, writing, and public speaking. She has written acclaimed books–including *Last Oasis,* published in eight languages and the basis for a 1997 PBS documentary–as well as more than 100 articles for popular and scholarly publications, including *Science, Foreign Policy, Scientific American,* and *Natural History*. Her article "Troubled Waters" was selected for the 2001 edition of *Best American Science and Nature Writing*. Between 2000 and 2008, Postel served as visiting senior lecturer at Mount Holyoke College, and then director of the College's Center for the Environment. Postel has received a Pew Scholar's Award in Conservation and the Environment and two honorary doctor of science degrees, and in 2002 was named one of the "Scientific American 50," an award recognizing contributions to science and technology.

ON A MONDAY AFTERNOON IN JULY 2009, I was sitting at my desk sorting through emails, articles, and reports when I saw a news headline that stopped me in my tracks. It was out of Paris, from Agence France-Presse, and datelined July 10: "Eastern Aral Sea has shrunk by 80 percent since 2006." The European Space Agency had just released an overlay of satellite photographs showing that between July 1, 2006, and July 6, 2009, the eastern lobe of the Aral Sea in Central Asia—once the world's fourth largest inland water body—had lost four-fifths of its water within just the previous three years. The news report said that it was now likely that all but a small remnant of the sea would disappear by 2020.

The memories came flooding back. On a fact-finding trip to the Aral Sea Basin in March 1995, I'd gazed out from a lakeside bluff on the outskirts of Muynoq, an old Aral Sea port town, but saw no water—just miles and miles of desiccated earth. The coastline was 25 miles away. Before me was a graveyard of ships, rotting and rusting in the dried-up seabed. Salt dusted the landscape like newly fallen snow. Toxic dust storms emanating from the exposed lake bottom would, on windy days, make the air hazardous to breathe and poison the surrounding land.

The societal landscape around me mirrored the physical one. Sixty thousand fishing jobs had been wiped out, and thousands of people had left the area. Muynoq resembled a ghost town. The people who remained in the "disaster zone" suffered from startlingly high rates of anemia, respiratory ailments, and a variety of cancers. Infant mortality was high. Never before had I grasped so viscerally the connections between the health of an ecosystem and that of the economy, community, and people who depend on that ecosystem.

But the most heart-wrenching memories from that trip were of a meeting in the city of Nukus, the capital of Karakalpakiya, an autonomous region within the newly independent country of Uzbekistan, a former republic of the Soviet Union. The people of Karakalpakiya were among those affected worst by the sea's tragic shrinking. We met with environmentalists, health specialists, and representatives of local groups working daily to make life better in the disaster zone. They spoke of the sea's demise as an unthinkable loss of their way of life and their cultural identity. They spoke of illness, death, degradation, and sadness over the disappearance of their beloved Aral Sea. I listened.

Then they asked us, the visiting "experts" from the other side of the world, to offer our ideas and recommendations. A wave of unease arose in my body. I was keenly aware that anything close to a total repair of the Aral Sea's destruction was impossible. What the regional governments and international agencies were likely to offer paled in comparison with what the region and its people needed. As my turn to speak approached, I felt paralyzed by an unsettling mix of sadness and panic. I looked around the room at the dozens of people who had so passionately expressed their concerns, their needs, and their desperation. What could I possibly say that would make a difference?

What came to me, moments before I was to speak, was the story of California's Mono Lake, which had some parallels with the Aral Sea story but had a happy ending. I had visited Mono Lake only once, in 1982. At that time, after four decades of Los Angeles's diversions of its tributaries, the lake was near its lowest recorded level. It had lost half its water, and its salinity levels had doubled. The haunting beauty of the lake, nestled on the eastern side of the Sierra Nevada, had made a deep impression on me. Its unique ecosystem supported brine shrimp, alkali flies, a variety of wildlife species, and millions of migratory birds. Although it didn't come close to supporting the fisheries and jobs that the Aral Sea did, it too was a treasure. Over the years I had carried its image and followed its story.

Just a year before I traveled to the Aral Sea region, the California State Water Resources Control Board had ordered Los Angeles to halt its diversions from Mono Lake's tributaries until the lake rose to a surface elevation of 6,392 feet—the level scientists had determined was necessary to restore and maintain the lake's health. Los Angeles would have to turn to conservation and other methods to meet its water needs. Combined with earlier court decisions on Mono Lake, it was a stunning environmental victory.

And so there in the poor town of Nukus, in the Aral Sea Basin, I told the Mono Lake story with all its David and Goliath–like qualities. I told how a small but extraordinarily dedicated group of people called the Mono Lake Committee had fought for the lake and, seemingly against all odds, had won. When I finished, I wondered if my intended message of hope had come through in the translation. As I saw smiles break out across the room, I realized that it had.

As I headed back home across the Atlantic, I knew deep in my bones that the Aral Sea would not have a happy ending like Mono Lake. The cards were stacked too much against it. The water diverted from the Amu Darya and the Syr Darya, the two rivers that feed the Aral, supplied irrigated cotton farms that were the backbone of the region's economy. Decades before, Soviet central planners had calculated that the rivers' water was worth more irrigating cotton in the desert than left in their channels to flow into the sea. The sea's demise had been planned, and neither the loss of the ecosystem itself nor the fate of the people affected had been given much weight.

Over the years, I continued to track the Aral Sea's shrinkage. River flows into the sea varied from one year to the next but averaged just 10 percent of pre-diversion levels—far too little to compensate for the high evaporation rates in the Central Asian desert. By 2005, the Aral had lost 80 percent of its water. That year, with international assistance, engineers constructed a dike between the Small Aral in the north and the Large Aral in the south, where the sea had divided in two 16 years before. The idea for the dike, which was 20 feet high and 8 miles long, was to salvage a remnant of the sea by preventing the smaller northern part from draining completely. It seemed like a good idea, even a cause for celebration to some in the region, as fisheries and wetlands in the little lake began to bounce back. But the larger lake, the original fabled Aral Sea, would be consigned to death.

So I knew the day would come when the Aral Sea was no more. I just never imagined, until the news out of Paris in mid-July 2009, that it would happen so fast.

It was in 1994, leading up to my trip to the Aral Sea, that I systematically looked at data on river flows from around the world. It was not just the Amu Darya and Syr Darya that were running low before they reached their final destinations, but many others as well—the Indus and Ganges in South Asia, the Nile in northeast Africa, the Jordan in the Middle East, the Colorado in the western United States, the Murray-Darling in Australia, and the Yellow in China. These were major rivers, collectively supplying hundreds of millions of people and millions of acres of irrigated land. I wrote up my findings for a magazine piece titled "Where Have All the Rivers Gone?"

Gradually, the ramifications of what was going on around the world sank in more deeply. During my research travels, I'd crossed the Yellow River by train in 1988, and had been stunned to see how little water was in the channel of what the Chinese called their "mother river"—the nation's second largest and the cradle of Chinese civilization. It was broad, shallow, and silty, and still had many miles to go before reaching the sea. The Yellow had been running dry in its lower reaches on and off since 1972, and the frequency and length of dryness was increasing as more and more water was diverted for irrigation and populations upstream. A few years later, in 1992, I'd seen the delta of the Nile River, where the small volume of fresh water flowing in no longer delivered enough sediment to replenish the delta, a rich agricultural region, which was sinking and being overtaken by the Mediterranean Sea.

But it was in the spring of 1996, a little over a year after I'd returned from the Aral Sea region, that I was once again brought up short by the human face of ecological degradation. This time it was closer to home, in the delta of the Colorado River, and the story was written in the deeply lined face of an elder member of the Cucapá Indian community in northern Mexico.

A proud, handsome people, the Cucapá have fished and farmed in the delta of the Colorado for perhaps 2,000 years. They harvested a grain they called *nipa,* a salt-loving plant known to botanists as *Distichlis palmeri* that tastes much like wild rice. It was a dietary staple. Fish was so abundant the Cucapá sometimes ate it three times a day. Known as people of the river, the Cucapá keyed their life to the Colorado's seasonal floods. Historical accounts suggest that 400 years ago as many as 5,000 Cucapá were living in the delta.

When I visited their community of El Mayor in May 1996, they were a culture on the brink of extinction. The Cucapá were to the depleted Colorado River what the Karakalpaks were to the shrunken Aral Sea: the voiceless victims of water management driven by dollar signs and power politics. Their way of life hung in the balance because the Colorado's water had been dammed and siphoned off to fill swimming pools in Los Angeles; to illuminate Las Vegas; and to irrigate crops in the deserts of Arizona, California, and the Mexicali Valley. Fishing and farming no longer sustained them. The annual floods that had naturally irrigated their nipa had disappeared with the construction of dams—Hoover, Davis, Parker, Palo Verde, Imperial, Morelos, and Glen Canyon—on the main stem upstream. The young tribal members were heading to Yuma and other cities to find jobs. Only about 50 Cucapá families remained south of the border.

One morning during my research trip, we took a low-altitude flight over the delta region to get a bird's-eye view of the landscape. I saw the dry channel of the Colorado trace the river's meandering path toward the Sea of Cortés, known north of the border as the Gulf of California. No longer was the sea getting sufficient nutrients to sustain its fisheries. No longer was the delta getting replenished by water and sediment. I looked down on mudflats, salt flats, and scattered, murky pools, and I thought to myself: Aldo Leopold must be turning in his grave.

In his classic work *A Sand County Almanac,* the great naturalist Leopold had written about canoeing through the delta with his brother Carl in 1922. He described the delta as "a milk and honey wilderness" and a land of "a hundred green lagoons." He saw deer, quail, raccoon, bobcat, and vast fleets of waterfowl. It was hard to believe the area below me could at one time have been the biologically rich and verdant place Leopold wrote about so eloquently.

Ironically, 1922 was also the year the river's fate was sealed. That year, representatives from the seven U.S. states in the Colorado watershed met at Bishop's Lodge outside of Santa Fe, New Mexico, and divvied up the waters of the Colorado River. During the years leading up to this historic gathering, considerably more rain and snow had fallen in the watershed than the long-term historic average. So the treaty signed by the seven states (an agreement with Mexico was signed later, in 1944) promised more water to the parties than the river typically carried. Moreover, no water was set aside for the river itself, or for the delta and Cucapá communities at the river's tail end. Engineers got busy building the dams and reservoirs needed to carry out the terms of the treaty. The first, Hoover Dam, was completed in 1935. The last, Glen Canyon Dam, was completed in 1963, after which the Colorado ran dry in its lower reaches for two decades.

So after my experience the year before in Nukus, I held my tongue when a Cucapá elder said with conviction: "I hope one day to see the river rise again." I did not intend to repeat what I had done in Nukus—tell a true story, but one that, given the circumstances, delivered a message of false hope.

Late in the afternoon of July 13, 2009, a few hours after I'd seen the jaw-dropping headline about the Aral Sea, I headed to the Albuquerque airport for a flight to Las Vegas. I was scheduled to meet a British film crew there for a shoot on Lake Mead, the vast reservoir on the Colorado River created by Hoover Dam. I tried to ignore, but couldn't, the hypocrisy of taking a carbon-intensive journey to a shrinking lake to talk about climate impacts on fresh water.

To prepare for the trip, I read up on some projections about the river and the reservoirs under different climate scenarios. The one that most caught my attention was by climate scientists Tim Barnett and David Pierce at the Scripps Institution of Oceanography at the University of California in San Diego. They had found that if the climate changes as expected, there is a 50 percent chance that Lake Mead, a water source for tens of millions of people and one million acres of irrigated land in the U.S. Southwest, will be dry by 2021. They estimated that the Colorado River system is losing, on balance, about one million acre-feet of water a year—enough to meet the demands of eight million people. Without more precipitation or a curtailment in water use, that net deficit will build year after year, and the volume of water in Lake Mead will keep shrinking until it is dry. The year 2021 had a distant ring to it, so I solidified in my mind that this unthinkable situation could occur in just a little over a decade.

Lake Mead is not only the granddaddy of reservoirs on the Colorado, it is the largest man-made reservoir in the United

States. When full, it holds about as much water as the Colorado River carries over a two-year period. With such a storage capacity, it ensures a steady supply of water for parts of Arizona, Nevada, California, and northern Mexico. Without that storage, the reliability of those supplies would diminish greatly.

The gamble in letting the water deficit build—and the lake level drop—is that the rains and snows will return sufficiently to replenish the Colorado River system. But the climate models suggest that this is less and less likely. If Lake Mead becomes dry within a decade or two, can the Southwest remain habitable in the way we now know it?

As I pondered this question, my view out the right side of the airplane as we headed into Las Vegas was of the water body itself—Lake Mead, there in all its glory, with a thick, white bathtub ring around its perimeter. The lake's level had dropped 100 feet since the year 2000. If it falls much farther, the Las Vegas valley's primary water intake will become inoperable. Because the lake is wider at the surface than at greater depths, like a cone-shaped coffee filter, the lake's drop in level represents a much bigger loss of water than it appears to. Although the lake's level was now only 10 percent below the 1983 high-water mark (which occurred after a strong El Niño weather event had delivered unusual amounts of precipitation to the watershed), the volume of water in the lake had dropped from 96 percent of capacity to 43 percent in just the last nine years.

As the plane lowered its landing gear, I strained my neck to watch the edge of Lake Mead recede. A few seconds later, a lush, green golf course came into view. With late afternoon temperatures pushing 107 degrees, there was scarcely a golfer in sight. Downtown, on the Strip, the dice-rolling gamblers had no idea how high the stakes were becoming.

Reasonable people can disagree about the details, but it's clear that the natural world around us is changing much faster than anyone, including scientists, would have imagined possible even five years ago. Glaciers and ice caps are melting, seas are rising, rivers and lakes are drying up, and floods and droughts in rich and poor countries alike are wreaking havoc with agriculture. So in the midst of these trends, one can be forgiven for wondering, is there hope for a water-secure world? Not just any hope, but *honest* hope. Is it really possible to join the Cucapá elder in hoping for a future in which the depleted Colorado—and all the rivers like it—will rise again to sustain the communities, both natural and human, that depend on them?

At this moment, we as a society are like the frog that chooses to stay in a warming pot of water as the heat is gradually turned up—unable to grasp the dire consequences of incremental change. Inch by inch, the water tables drop. Mile by mile, the rivers run dry. The trends are not good. Yet we stay the course, refusing to recognize that, for safety's sake—for survival itself—a big change is necessary. We pretend not to know. Denial, as has been said, is not just a river in Egypt. It flows in every one of us.

But there comes a point where denial stops offering its false comfort, and we must look reality squarely in the eye. And if this moment arrives in time, before irreversible thresholds are crossed, we get to the root of honest hope. It lies, I think, in this idea: As the unthinkable begins to happen, actions we thought impossible become possible.

In fact, when I flip my water lens from the panoramic view and zoom in to take a close-up, more localized look, I see something extraordinary: Little shoots of honest hope are sprouting all around. Community farms and backyard and rooftop gardens are popping up, reducing both the energy and water demands of our diets. Farmers are shifting from flood irrigation to highly efficient drip systems, which deliver water directly to the roots of plants. Some are doubling or tripling the amount of crop per drop they get from their limited water supplies. Cities such as Boston, San Antonio, Seattle, Los Angeles, and New York are driving down urban water demand by promoting conservation in homes and businesses.

More and more communities are taking the "waste" out of wastewater by finding ways to recycle and reuse it. Some forward-thinking corporations are promoting more sustainable water use throughout their supply chains—from farm to factory—to shrink the water footprint of their products. Growing numbers of obsolete or harmful dams are being dismantled and removed—including some 430 in the United States alone over the last decade—allowing fish to return to native spawning grounds and rivers to flow more naturally again.

Even some national governments have taken bold steps forward. South Africa has established a Water Reserve, declaring that the highest priority for water use in the country is to meet the basic water needs of all people and ecosystems in order to sustain human and ecosystem health. In Costa Rica, the government compensates landowners in forested watersheds for conserving rather than cutting down their forests in order to safeguard the array of societal benefits—from clean, reliable water supplies to species diversity and carbon sequestration—that healthy watersheds provide. And in North America, the eight U.S. states and two Canadian provinces that surround the Great Lakes have agreed to ban diversions of water out of the Great Lakes Basin, essentially saying that there is no surplus water there; it is all needed to ensure the future health of the Great Lakes.

When I take such examples and replicate them many times, I can arrive at a vision of a world in harmony with the planet's life-giving water cycle. It's a world in which we'd see more rivers actually flowing like rivers again, one in which farmers would use half as much water to provide nutritious diets to the world, and one in which computers and clothes would be made in factories that treated and reused all of their process water. It's one in which ecological engineers would help communities rely on nature's solar-powered infrastructure—wetlands, floodplains, and forested watersheds—to purify water, help control floods, and recharge groundwater, all while preserving habitats for fish and wildlife and beautiful places for people to enjoy. It's one in which the water productivity of national economies—the volume of water used to produce a dollar of gross national product—would be up to ten times higher than it is today. Overall, it is one in which water management would be less about pipes, pumps, and pouring concrete, and more about ideas, innovation, and ecological intelligence.

This is a world most of us would want to live in, whereas the world we will create if current trends go unaltered is not. The leap we need to make to get to this better place—to exit the pot of water before it's too late—must be big and bold. It will require that we shake up the old rules of the water game, crafted when water seemed abundant, maybe even inexhaustible, and replace them with new ones more aligned with the world of today. Pricing, markets, regulations, rights, incentives, legal constructs—all of these tools will need an overhaul for the leap to be big enough to get us out of the danger zone.

But while necessary, these changes alone will not be sufficient. Some fundamental element is missing. And it has to do with what I saw on February 13, 2004, as I strolled around—of all places—the Charlotte, North Carolina, airport.

My connecting flight to San Diego was delayed by several hours. I grabbed some lunch, wandered through a bookstore, and went over my notes for the next morning's talk. I was scheduled to speak about global water issues to a conference of investors gathered in Carlsbad, just north of San Diego. Water was only a small part of the conference agenda, but I anticipated some questions about investment opportunities in desalination—the process of creating fresh water by desalting seawater. A desalination plant was slated to be built, right there in Carlsbad, but more than that, desalination has that silver-bullet quality that investors often find attractive. After all, wouldn't the ability to remove salt from seawater on a very large scale just about solve the world's water shortages?

Yes, but—I needed to sharpen my response. For sure, it's a lifeline for energy-rich Persian Gulf countries that can afford to turn oil into water, but most of the world can't afford to do that. Desalination produces not only drinking water, but also a toxic briny waste product. More to the point, why use climate-altering fossil fuels to produce water when conservation and efficiency investments can meet new water demands at less cost and without the environmental downsides? And isn't it counterproductive to droughtproof drinking water supplies with a technology that spews more greenhouse gases into the atmosphere and thereby increases the risk of drought?

Anyway, I was working all of this out in my head as I wandered through the Charlotte airport. With a couple more hours to kill, I decided to step outside for some fresh air. I walked around a parking lot, enjoyed the Friday afternoon sun, and then headed back toward the terminal. And that's when I saw what I'd never seen so clearly before. It was personified, right there, in a little girl, barely two years old, who, upon rounding a corner and seeing a beautiful fountain of water, ran shrieking in sheer delight to be near it, to touch it, to feel it, to bond with it. The pull appeared as strong as an electron toward a proton.

We are connected. All the water here on Earth now is all the water there ever was, and ever will be. Through the cycling of water, across space and time, we are linked to all of life. My morning coffee might contain water that the dinosaurs drank. Earth's water, embedded with the wisdom of the ages, is literally in our blood. And as molecules of water circulate from sea to air to land—through the clouds, through the rivers, through the trees, through the frogs and fish and mussels and beetles and ants and birds and bees and everything alive, now and then and yet to be—we are connected. And this is what I think the Cucapá elder knew but what I had only begun to grasp: that we cannot sever one part of the Earth from another without damaging the whole.

Deep in our bones, we know this. We have just pushed that knowing out of the way.

But as the unthinkable keeps happening—as water disappears from rivers and lakes and the aquifers beneath our feet—I believe we will begin to awaken to our kinship with water, which springs from knowing at a cellular level that water's gift is life. No water, no life. Our innate love of life will rise up and call upon us, individually and in our communities and beyond, to do something with water we have so far found impossible to do: We will share it. We will share it because water's gift is life, and we are connected to all life. We will start to live by a new ethic that says all living things must get enough water to survive before some of us get more than enough.

That's a big leap from the me-first thinking that determines most of our decisions about water today. But we need to make it—now.

RESPECT FOR WATER

Fred Pearce

Fred Pearce is a freelance author and journalist based in London. He has reported on environment, science, and development issues from 64 countries over the past 20 years. Trained as a geographer, he has been environment consultant to *New Scientist* magazine since 1992, with a special interest in water issues and climate change. He also writes regularly for the *Guardian* and other British newspapers. Pearce's books, which have been translated into at least 12 languages, include *When the Rivers Run Dry, Confessions of an Eco Sinner, Earth: Then and Now, With Speed and Violence* (on climate change), *Keepers of the Spring,* and *Deep Jungle.* His next book explores population and the environment in the 21st century.

FOR MOST OF US, water comes out of the tap, or from bottles. But that is the boring bit. My lifelong voyage of discovery has been about flowing water—where it comes from, where it goes, and how we get our hands on this slippery, and remarkably heavy, substance. Nothing is more precious to us, more essential, than water. Yet, paradoxically, we treat it with disdain and an extraordinary lack of care. Now that there are almost seven billion of us on the planet, we have to change.

I think my voyage started when I was a boy, about ten years old. I lived in the countryside of rural Kent—"the garden of England." For an end-of-term school project, a couple of us were told to survey our local river, which began at a spring about a mile from the school gate. It was one of two streams that started in my village: The Stour went east; the Len went west. We went west, just for a couple of miles, through pools and eddies and woodlands and past fields and backyards. We wrote up our report afterward. It was enough. I've followed rivers ever since, and written up reports. They were the subject of my first book, and my fifth and eighth and, well, you get the picture.

What I have found extraordinary is how few truly wild rivers there are left in the world. How few undammed, unembanked, untapped rivers there are, that just run out of the hills through their floodplains and down to the sea. We rightly rage against the loss of the rain forests. But "old man river" isn't what he used to be either. Sadly, we mourn his passing far less.

The Yellow River in China, along which the world's longest lasting civilization has prospered for some 4,000 years, is called China's "joy and sorrow." And our relationship with many rivers fits that description.

Looking back on a lifetime of river-watching, I remember the joy the most. Early one January day back in the 1990s, I remember going to France, to a town called Le Puy, near a gorge high on France's longest river, the beautiful Loire. The "Loire Sauvage," they called it. I spent the morning in the gorge with a musician who gave me a copy of his new disc in which he sang about the river, and scrambled down the gorge to fill a bottle with its bright water. I spent the afternoon at a campsite where a handful of French youths had stuck it out through a long, cold winter, protecting the site where engineers had tried to start construction of a dam to barricade the gorge and tame the river.

Then, as the sun set, I went into town to buy some buttons at the protesters' campaign headquarters, only to discover with delight that the government had unexpectedly given in. The dam would not be built. A night of red wine, goat cheese, and Gallic emotion followed down in the gorge. A girl who had kept vigil with the boys through the cold winter danced with an elderly Gauloise-smoking peasant farmer. The river had been saved.

Saving rivers excites huge passions—passions that can topple governments. Back in early 1989, I visited campaigners in communist Hungary who were trying to prevent construction of the Nagymaros Dam on the River Danube at a beauty spot near Budapest called the Danube Bend. It was the site of the country's ancient capital. The underground movement against the dam had been started and run by a retiring biologist named Janos Vargha, who told me he believed it would kill the river. By 1989, the campaign had broken cover to become the first open defiance of the Soviet-backed government, bringing tens of thousands of people out onto the streets to face tear gas and truncheons.

Popular opposition to the dam grew so great—and the determination of the communist regime to press ahead so strong—that the campaign eventually brought down the government. Reformist ministers who came into power during 1989 canceled the project, and then, emboldened, they opened the border between Hungary and the West.

The world held its breath. Within weeks, thousands of East Germans were taking their "holidays" in Hungary and crossing into Austria and West Germany. The Iron Curtain between west and east had been breached. Within weeks, they were tearing down the Berlin Wall. It had become redundant. Looking back, it seems to me that it wasn't Ronald Reagan or Margaret Thatcher or even Russian leader Mikhail Gorbachev who brought down the Berlin Wall. It was a Budapest biologist. By opposing the dam, he opened the political floodgates.

When I met him recently in a Budapest coffee bar, his once fierce beard had gone gray. But when I suggested his role in one of the greatest political events of the last half century, he smiled the quiet, unassuming but determined smile I remembered from 20 years before. Yes, he supposed I was right. But he acted as he did not for politics, but to save a river.

And he was right. The few wild rivers that remain, with their rapids and tangled deltas and unruly floods, are a living lesson on what most of the world's rivers have lost. One of my most vivid journeys was to the Mekong in Southeast Asia—where I found a river that flows backward.

The Mekong still has its huge annual flood, driven by the monsoon rains and the melting snows at the river's source in Tibet. Its flow increases 50-fold during the summer flood season. No river on Earth today has such a variation in flow. In fact, so much water flows downstream through China and Burma into Thailand, Laos, Cambodia, and Vietnam that the river bursts its banks everywhere. And in front of the royal palace at Phnom Penh, the capital of Cambodia, so much water is coming down the main river that it pushes up into a tributary called the Tonle Sap, which goes into reverse.

From June to November each year, the Tonle Sap flows uphill for 120 miles, forming a huge lake in the forests of central Cambodia. This flooded forest has, from time immemorial, been a massive breeding ground for fish. In the silty water among the tree roots, billions of fish fry grow into fat adults. The flooded forest is one of the most productive ecosystems on Earth. And each November, as the flood abates, the Tonle Sap turns, the lake empties, and the fish swim out. Over the coming months, they migrate for thousands of miles up and down the Mekong—filling nets that feed tens of millions of people.

Two-thirds of the fish in the Mekong begin their life in the Tonle Sap. This strange river reversal makes the Mekong the second most productive inland fishery on the planet, exceeded only by the Amazon. It means that the people of Cambodia, among the world's poorest, are among the best fed.

This all happens under the eye of the great temples of Angkor Wat. These remains of an ancient jungle civilization that prospered a thousand years ago sit on the north shore of the flooded forest. It was the Mekong fishery, centered on the lake, that sustained this empire. Even today, the lake is crowded with floating fishing villages that cruise the waters, catching fish and farming crocodiles. To visit them is to visit a different world, where water is king and land is almost forgotten.

The most extraordinary product of this fishery is the giant catfish, one of the wonders of the riverine world. The protected species grows up to ten feet long and can weigh a third of a ton. Its numbers are declining, but it still lurks in huge hollowed-out pools on the riverbed, and occasionally turns up in the nets that fishermen put across the Tonle Sap. I met an American fish scientist, Zeb Hogan, who slept by his mobile phone each night waiting for a call from the fishermen, at which point he would rush out and measure and tag a beast before they put it back into the river.

But this riverine cornucopia will soon come to an end. Kept away by nearly half a century of war in Indochina, the dam builders are finally arriving. They want to tap the monsoon flood behind dams and generate hydroelectricity by letting the water out, gradually, through turbines.

In October 2008, Chinese engineers finished construction of the Xiaowan Dam in a gorge in the upper reaches of the River Mekong. At 900 feet, it is the world's tallest dam, as high as the Eiffel Tower. The reservoir will eventually be more than 100 miles long. It will help keep the lights on in Shanghai, 1,200 miles to the east. As China rushes to industrialize, a total of eight hydroelectric dams are planned on the Mekong. By 2014, engineers will have completed the Nuozhadu Dam, which will have an even larger reservoir.

The Mekong is destined to become China's new water tower and electrical powerhouse. The cascade of dams will be able to store half the entire flow of the Mekong. The river will henceforth rise and fall at the whim of engineers rather than nature. In May 2009, a report from the UN Environment Programme warned that these dams were "the single greatest threat" to the future of the river and its fecundity. The reversal of the Tonle Sap will end, and with it the river's fisheries.

We should weep for the Mekong. But what is happening here is only what has happened already on countless rivers round the world. The Columbia River in the American Northwest is now barricaded by the Grand Coulee hydroelectric dam and others. The Grand Coulee is the seventh largest hydroelectric dam in the world. Built in the 1930s, the Grand Coulee and the lights it turned on were regarded as a modern miracle. Woody Guthrie wrote a song in its praise. Hydroelectricity is a green and renewable source of energy. But the Columbia was once one of the world's great salmon rivers, and economists now say that the salmon catch lost each year would be worth more today than the electricity generated by the dam.

Rivers are also abused to obtain water to irrigate crops. Round the world, farms take two-thirds of the water that we appropriate from nature—much more than industry or domestic use. The amount of water each of us needs to get us through the day is staggering.

I was amazed when I tracked my own water consumption to work out my "water footprint." On average, I drink not much more than a gallon of the stuff per day. Even after washing and flushing the toilet, I get through at most 40 or 50 gallons. But that is just the start. It is only when I add in the water needed to grow what I eat and drink that the numbers really begin to soar.

It takes between 250 and 650 gallons of water to grow a pound of rice, more water than many households use in a week. For just a bag of rice. It takes 130 gallons to grow a pound of wheat, and 65 gallons for a pound of potatoes. It takes 3,000 gallons to grow the feed for enough cow to make one quarter-pound hamburger, and between 500 and 1,000 gallons for that cow to fill its udders with a quart of milk.

I turned those stats into meal portions and came up with more than 25 gallons for a portion of rice, 40 gallons for the bread in a sandwich or a serving of toast, 130 gallons for a two-egg omelette or a mixed salad, 265 gallons for a glass of milk, 400 gallons for an ice cream, 530 gallons for a pork chop. Every teaspoonful of sugar in my coffee requires 50 cups of water to grow it. Which is a lot, but not as much as the 37 gallons of water (or 590 cups) needed to grow the coffee itself. My pint of beer requires another 66 gallons or so to grow the barley and hops and process them into my favorite brew, and a glass of brandy after dinner takes a staggering 530 gallons. Even that is not the end. I worked out that I could fill my bathtub 25 times over with the water needed to grow enough cotton to make one T-shirt.

I reckon that, as a typical meat-eating, beer-swilling, milk-guzzling, cotton-wearing Westerner, I consume in this way as much as a hundred times my own weight in water every day. Growing the crops to sustain me for a year must take rather more than half the contents of an Olympic-size swimming pool—about 2,000 cubic yards. (Hold that number; I will come back to it.)

Water used to be thought of as the great unlimited resource. And the water in our rivers is indeed renewable—moving water that nature constantly recycles as it flows to the oceans, evaporates into the air, and returns to the Earth in new, clean rain. This natural water cycle is still raining the water in which the dinosaurs bathed and the first fish swam. It is the basis of life on Earth, and of our civilizations.

But only so much water runs through this cycle each year. There is only so much that we can tap, even if we return every drop to the rivers or oceans after use. Hydrologists estimate that about 3,400 cubic miles of water make the journey down rivers to the sea every year. It's a crazy unit, the cubic mile, but I don't know of a better one. At any rate, that is the renewable water that, theoretically, we can tap for our uses.

Unfortunately, many of the world's greatest rivers are in regions where few people can or want to live. The three rivers with the biggest flows—the Amazon, the Congo, and the Orinoco—all pass through inhospitable jungle. And two more of the top ten run mostly through Arctic wastes. Take out these and we are left—for practical purposes and with current engineering technology—with around 2,200 cubic miles of river water for our needs. (OK, we could set up a global network of water pipes, but if you have picked up a bucket of water recently, you'll know that it is heavy stuff. As soon as you start pumping it uphill, the whole business quickly becomes prohibitively expensive.)

That 2,200 cubic miles of water still amounts to about 1,500 cubic yards a year for every citizen on the planet. It doesn't sound bad. But remember I calculated my own annual water use at about 2,000 cubic yards a year. I imagine most of the world would like to live as well as me. So, supply, 1,500 cubic yards a year; demand, rather more. Amazingly, we have a problem. We are coming up against real limits to what until recently seemed an inexhaustible resource.

And that is why we are seeing great rivers like the Indus in Pakistan and the Nile in Egypt, the Yellow in China, the Murray in Australia, and even sometimes the Rio Grande and the Colorado in the United States and Mexico, all running dry for parts of the year. Literally all their water is being taken by humans. In the dry parts of the world—where the rivers are smallest—most of that water is going to irrigate crops that would otherwise shrivel in the fields.

I remember standing one day on the bridge over the once mighty Rio Grande in El Paso, where the river forms the border between the United States and Mexico. I watched the river flow along its concrete, barbed-wire, flanked channel below. It was, I noticed, flowing upstream. There was so little water in the channel that the wind coming up from the Gulf of Mexico was sufficient to blow the great river backward.

Later, I stood on the banks of the Rio Grande a couple of hundred miles downstream in a small town called Presidio, where the fields have been irrigated for more than 400 years, probably longer than anywhere else in the United States. Now farmers are going bankrupt and the fields are returning to sagebrush and salt cedar because there is no water left in the river most of the way from El Paso to Presidio.

In the Alamo coffeehouse, local farmer Terry Bishop told me he was selling up. His land was beside the river, and he had a legal right to take 8,000 acre-feet of water a year from it, enough to flood all his fields to more than a yard depth. But what value was that right when there was no water in the river?

You won't spot this in any atlases, but, in fact, the mighty Rio Grande is now two rivers—an impressive example of how our use of water is changing the geography of the planet. The main U.S. arm, rushing out of the Rockies, gives out at El Paso, 600 miles from the Gulf of Mexico. Its bed is then dry for 200 miles until it is replenished from Mexico just past Presidio.

Water, much more than the gun, made the American West. But today the American West is running short of water. Cities like El Paso are buying up huge areas of farmland, not because they want to farm but because they want the rights to the water beneath. As a result, these underground water reserves are disappearing. America has pumped almost dry the great Ogallala Aquifer beneath the High Plains. It was done in a good cause: to provide irrigation water and restore farming to the areas that were all but destroyed by the Dust Bowl drought of the 1930s. But now what was once one of the largest underground water reserves in the world, running from South Dakota to Texas, is close to empty.

The same thing is happening in India, with consequences that could create a dust bowl that kills millions. A century ago, British colonial engineers in India built the largest network of irrigation canals in the world to grow food and cotton for the British Empire. Today, they feed a billion Indians. But there is no water for many of those canals, because the rivers are dry for most of the year. Virtually the only water in India outside the short monsoon season is underground. Indian farmers are some of the most innovative in the world. In the past decade, more than 20 million of them have bought drills to tap that water, and cheap Yamaha pumps to bring it to the surface and water their crops. As a result, I visited villages where water tables were until recently only a few feet from the surface but are now hundreds of feet down.

Thanks to this water, India, which 40 years ago was a center of famine, today feeds itself—but at great cost. Farmers are taking from underground 30 cubic miles more of water every year than the rains replace.

India is living on borrowed water and borrowed time. The crash will come, not everywhere at the same time, but inexorably nonetheless. And dozens of other countries in Asia are going the same way as their rivers run dry.

In the process, new deserts are being created. This is most advanced—and certainly is happening on the largest scale—in Central Asia. I took a long journey through the former Soviet republic of Uzbekistan, all the way from the capital Tashkent to the shores of the Aral Sea, reputedly the fourth largest inland sea in the world. It took two days' driving, the entire route through cotton fields. Uzbekistan is one of the world's largest producers of cotton, thanks to a vast irrigation network set up in the desert by the Russians in the 1960s and 1970s. The cotton once clothed the Red Army. But since the breakup of the Soviet Union—since Commissar Cotton went home—the country's new capitalist rulers have become fabulously rich by running a cartel selling the country's cotton to the world.

It was only when I got to where the Aral Sea should be that I saw the horrific consequences. The town of Muynoq, where I put up at the only hotel, was once a bustling seaside resort and fishing port. Its trawlers plied the water of an inland sea the size of Belgium and the Netherlands combined, but no longer. I found old ships beached in the harbor and a huge fish-canning plant abandoned, its corridors still decorated by pictures of heroic Soviet trawlermen.

From the old promenade, where holidaymakers from as far away as Moscow once took cruises, I looked out to sea. It was sand and sand and sand. I took a drive. About three miles out, I spotted a fox, trotting through the tamarisk. But I never saw the sea. Satellite images show its shoreline is now 100 miles north of Muynoq—three small, fishless, hypersaline sumps in a new desert, entirely unmapped and largely unvisited.

This has happened because, for the past 50 years, almost the entire flow of two giant rivers, each the size of the River Nile, has been removed to irrigate the cotton fields of Uzbekistan and neighboring Turkmenistan. Without the inflow from those rivers, the Aral Sea has simply been evaporating in the sun.

But this is not even a straight swap: cotton for fish. For the rivers bring salt as well as water, and over the years the salt in the irrigation water has poisoned the cotton fields. Now extra water is needed to flush the soils clean each spring before the cotton can be planted and the real irrigation begun. So more and more water is used, while less and less cotton is grown.

The UN calls the death of the Aral Sea the greatest environmental disaster of the 20th century, but it is an amazing human tragedy as well. The salt is poisoning everything and everyone. It is in the water; in the food; and in the air, whipped up in the sandstorms that blow off the dried-up seabed. Salt, even more than the agricultural chemicals that also blow in the sandstorms, is a killer. A local doctor, Oral Ataniyazova, told me life expectancy in the area round the dried-up seabed is now just 51 years. Some 97 percent of the 700,000 women in the region are anemic. "All our women are sick, and so are all our newborn," she told me. One in every twenty babies is born deformed. In some villages the mothers' milk is so salty babies will not suckle.

I talked about all this with a group of farmers in Chumbai, one of the towns near the old seashore. These aging, weather-beaten men in shabby jackets and pullovers told me how the salt was killing them. "You can see it in the faces of everyone living here," said the manager of a state farm, standing up to speak, taking off his hat and holding it gently in front of him. "We are all affected. There is a lack of good blood. All the women have it and it is worst in pregnancy. Lots of people die, especially the children."

He went silent for a few moments, red faced and holding tight to the brim of his hat. "My daughter and son were both in hospital for six months. Every family has someone like that." Nobody would say more. This was private grief from the victims of a fading Soviet dream and a dying sea.

There are more tragedies happening because of the rampant misuse of water round the world. In northern Nigeria I visited the inhabitants of the Hadejia-Nguru wetlands, a large oasis on the edge of the Sahara, where they fish and herd cattle on flooded pastures, gather reeds, and irrigate a few crops. But these wetlands were drying up because water from the rivers that fed them was being taken to irrigate a new government irrigation project near the city of Kano. The government said the project would "green the desert." In fact, it was creating desert.

The world is fixated on supplying ever more water but conserves and manages it with appalling abandon. Engineers spend billions of dollars building dams and irrigation projects but rarely consider how well that water is used. We are addicted to putting water in pipes and canals, where we can control it, rather than leaving it in rivers and wetlands where we may have less control but may gain more wealth. We can measure the electricity we generate from a dam and the crops grown in the irrigation scheme more easily than we can count the fish that were never born because a river was emptied or the flooded pasture lost.

And sometimes worse disasters happen. One of my most harrowing journeys was to Bangladesh. Back in the 1980s, doctors decided that surface water in this extremely wet country was too polluted, and it made people sick with diarrhea. So they proposed sinking millions of simple tube wells with hand pumps into backyards across the country to capture the underground water. Aid agencies like UNICEF began drilling. Private companies in

Bangladesh joined the movement. It was popular with the people, though some old-timers warned it was a bad idea. The underground water, they said, was "the devil's water."

Within a decade, some ten million tube wells had been sunk. Most people drank underground water. Then the strange cases of blotched skin and lesions and cancers began to emerge and increase to an epidemic. Finally, someone thought to test the underground water. They found that much of it contained arsenic at concentrations hundreds of times greater than accepted safe levels. Tens of millions of people were drinking poisoned water. It took years to build up a lethal dose, but people were starting to die.

I went out with a screening team to check the water in one of the country's 68,000 villages, called Dipordi. We chose it at random, but nobody had checked its tube wells before. Our chief checker, Akhtar Akhmad, arrived like a traveling medicine man. The villagers gathered to hear the results of the water from nine wells they had brought in bottles. One was off the scale at 50 times the World Health Organization's recommended limit of arsenic.

Then, they bared their bodies so Akhtar could check for telltale signs of poisoning. Symptoms were already showing after seven years of the new water (usually it takes ten years for symptoms to show). "This is the start of something big," said Akhtar. Before we left, he painted the poisoned pumps red and suggested everyone drink from the others.

But it was too late for a man we met in the next village. Abdul Kasem met us with a gruesome carcinoma on one hand and in the other a bottle of water, which he wanted tested. Sure enough, his water was lethal. The World Health Organization has called this epidemic "the worst mass poisoning of a population in history." Even though the slow process of finding the poisoned wells and identifying alternative safe ones in each village is a gargantuan task, it goes largely unreported. A lot of deaths still pass unnoticed out in the Bangladeshi countryside.

Perhaps the old-timers who warned about the "devil's water" were retelling ancient warnings handed down from past generations. Modern science tends to dismiss folk knowledge. But some of the most inspiring things I have seen on my journeys through the world of water have been efforts to reach back to a time past when we managed water much better.

In northern Syria I saw farmers who maintain underground tunnels known as *qanats,* dug centuries ago with extraordinary skill to tap water beneath the hills. Qanat construction is a technology seen at its most advanced in Iran, where some tunnels are many miles long. But most have dried up because modern boreholes and pumps have lowered the water tables too much. The wisdom of the ancients has been squandered.

In India, I went to villages where, as the rivers run dry and underground waters diminish, they are learning to catch the rains again. Before Westerners arrived and promoted building dams, the construction of low walls across the land to corral and store the rains, or divert them underground, was the traditional way of doing things. Now it is returning, promoted by ascetic supporters of nationalist and spiritual leader Mahatma Gandhi, Hindu religious groups, and even retired policemen.

I met Haradevsinh Hadeja, a cricketing ex-policeman in his village of Rajsamadhiya in the backwoods of the arid western Indian state of Gujarat. In the near desert, he and his fellow villagers had created a large lake by trapping the annual monsoon rains. The lake watered their crops, so they got three harvests a year.

Neighboring villages got one crop and often relied on government tankers even for drinking water during droughts. But not here. "We haven't had a water tanker come to the village for more than ten years," Hadeja told me. "There is no rain more than before. We just use it better. We don't let it wash away."

That, it seems to me, is the key lesson we must all learn. We will never have more water. Water may be recycled by nature, but it is also in a closed loop. There will never be more of it, or less. But to water a planet of seven billion people, we simply have to use it more sensibly, and waste less.

We have to value water for what it is, the font of our lives. A world away from the Indian desert at the University of Zaragoza, I met Pedro Arrojo-Agudo, a Spanish economist who had campaigned against a scheme to dam the River Ebro in wet northern Spain and send its waters south to fill taps and swimming pools in hotels and irrigation channels in the arid south.

The scheme was, he said, corrupt, expensive, mad, and completely unnecessary. There was enough water for all. What was needed, he said simply, was not more water but a "new water ethic," a new respect for water. He is right.

ANOTHER AFRICA IS POSSIBLE

Maude Barlow

Maude Barlow is the national chairperson of the Council of Canadians and senior adviser on water to the president of the UN General Assembly. She also chairs the board of Washington, D.C.–based Food & Water Watch and is a councillor with the Hamburg-based World Future Council. Barlow is the recipient of eight honorary doctorates as well as many awards, including the 2005 Right Livelihood Award (known as the Alternative Nobel Prize), the Citation of Lifetime Achievement at the 2008 Canadian Environment Awards, and the 2009 Earth Day Canada Outstanding Environmental Achievement Award. She is also the best-selling author or co-author of 16 books, including the recently released *Blue Covenant: The Global Water Crisis and the Coming Battle for the Right to Water.*

THE WORK OF A "WATER WARRIOR" can be very painful. I have become acutely aware of the inequitable access to water in different parts of the world, and even between social classes inside countries. Quite simply, if you have enough money, no matter where you live, you can have all the water you want for golf courses, swimming pools, water fountains, and fancy lawns. If you do not, you watch your children die of horrible waterborne diseases or hunger as parched lands lose the ability to grow food. My "water journeys" have taken me to many water-devastated parts of the world, but none more wrenching than Kenya in East Africa, now experiencing the worst drought in its history.

In January 2007, I attended the seventh World Social Forum in Nairobi along with a great team from my organization, including the Blue Planet Project's Anil Naidoo, and I was dazzled by the color, music, and vision of 65,000 delegates, mostly African, in search of our theme, "Another world is possible" (or in this case, "Another Africa is possible"). Access to clean water was understandably a central rallying cry for the participants, as the country is suffering from years of pollution, neglect, and mismanagement of its water supplies. Kenya, with a landmass twice the size of Great Britain, is everywhere parched. All of the lakes are polluted and overdrawn, and the groundwater supplies and rivers are disappearing. Whole towns have no water whatsoever, and it is not uncommon for women to walk six hours a day to find water for their families. Every day many children die from stinking, poisoned water.

I was deeply concerned that in order to provide water for such a huge gathering (oddly, held in a huge Chinese-built sport stadium) local wells had been drawn down and the adjacent communities were suffering from severe water restrictions. I was even more concerned to learn that this water was bottled and sold commercially, even to the activists at the forum. In my keynote speech, held in a big, dusty tent under a sweltering sun, I held up a bottle of water and said, "If another world is possible, then another World Social Forum without commercial water is possible!" For that, I got a great round of applause.

But for me, the most important work of that trip took place off-site. I went deep into the Nairobi slum called Kibera, where one million people live on 0.8 square mile of land. The average home is six square feet and has five people living in it. Half the population is HIV positive. Water is so scarce in this vast slum that there is not a patch of green anywhere, and local thugs sell drinking water for exorbitant prices out of the range of most of the residents. Local people are also charged money to use the

600 stinking, padlocked public pit latrines, so most defecate into plastic bags and throw them away—hence the term "flying toilets." I had a guide, of course; it would have been dangerous to enter Kibera alone. Alex, one of the porters from our hotel, who also lived in Kibera, led me through the alleys and shacks and introduced me to the local leaders struggling to meet the needs of their people.

Everywhere the children, many suffering from various eye diseases that made them appear blind, shyly watched or came up to touch me, always with a gentle smile, never begging. I left that terrible place (there are 200 similar slums in Nairobi alone) with the thought of my grandchildren back in Canada, their whole life spread out ahead of them with every opportunity, and my heart ached for these children, with so much potential but whose life will be cut short by the brutality of their environment.

The next day, Alex took a group of us, including Wenonah Hauter of Food & Water Watch and local biologist George Ogendi, deep into Kenya's Rift Valley to the haunting Lake Naivasha on a site visit. A local elder rowed us (standing up) in an ancient wooden boat onto the exquisite soft, blue waters of Lake Naivasha. The lake is home to one of the last major wild hippopotamus herds in East Africa, and we were eager to see them up close. Our guide, however, had a sensible fear of these beasts and steered a wide path around them as they sunbathed in the shallows and reeds. Lake Naivasha, a UN Ramsar (wetland protection) site, lies surrounded by volcanoes on the floor of the Great Rift Valley and is a paradise of biodiversity, flush with giraffes, zebras, water buffalo, lions, wildebeests, and at least 495 species of birds. I pointed to an island and declared, "That looks just like the movie *Out of Africa*!" Our gentle guide laughed and said that it looked like the film because that is exactly where it was made.

I could almost see Meryl Streep waiting for Robert Redford to land his plane and join her for "sundowners" on the veranda of her stately farmhouse.

George Ogendi and our guide then told us the sad tale of this lake, now on the brink of extinction. Until 1904, when the government signed an agreement opening it up to European settlers, Lake Naivasha and the land around it were protected to provide grazing and hunting for the Maasai people. Soon, European settlers had bought up all the best land, building plantations on the shoreline and then surrounding these homes with a ring of flower farms. The population started to grow and then exploded, from 7,000 in 1985 to more than 300,000 today, to service the flower industry. The vast majority of the workers—all black and mostly female—and their families live across the road from the farms in slums with no running water and pit latrines that leach into the lake. Our guide pointed out the one remaining public access site to the lake for the Maasai and explained that paid thugs discourage them from asserting their legal right to use this access point and fish the lake. The resulting illegal poaching had set up a brutal gangland-style struggle, one in which the famed nature filmmaker Joan Root got caught up. Our guide showed us the 88-acre home where Root was brutally murdered in the middle of the night by armed gang members in January 2006, a victim of turf warfare for this dying lake. Her murder has never been solved.

Kenya is the largest producer of cut flowers in Africa and the leading supplier to Europe. Britons alone spend three billion dollars a year on cut flowers, and Kenya has one-quarter of that market. About 30 major growers, almost two-thirds of them foreign owned, surround the lake with huge industrial farms, closed to the public with iron gates and armed guards. Roses are 90 percent water, and Europe is using Lake Naivasha and other African lakes to protect its own water sources from exploitation. The results are catastrophic: The lake is half the size it was 15 years ago, and the hippos are dying in the parched sun. If nothing is changed, we were told, the lake will be a "putrid puddle" in ten years.

We wanted to see a flower farm up close, but no owners would open their big, iron gates to let us in. Our guide knew one of the local workers, who told us to drive around to the back of his factory farm, where, glancing nervously around, he led us in by the workers' entrance. The first thing we saw was a sign over the entrance warning employees that their life would be in danger if they removed company property. We were then taken to the rows upon rows of roses thirstily sucking up the water from Lake Naivasha, which we filmed. The workers told us of terrible working conditions, including rampant sexual harassment of female employees and unprotected use of huge amounts of herbicides and pesticides. At the approach of a manager's car, our friends hustled us out the back way and we sped off in Alex's car.

Back at the World Social Forum, already feeling deep emotions from these visits and haunted by the faces of workers and children, I openly wept when 250 grassroots activists from 40 countries right across the continent held hands and danced to celebrate the newly formed African Water Network, the first pan-African network to fight for water justice for all Africans. Realizing the odds against these exceptional activists, some from the poorest communities on Earth, I could not help but understand the motto of the World Social Forum in a new way. Another world indeed was possible, but clearly, only if there was an international movement of solidarity to make it real. In moments such as this, I rediscover the human connection that makes me want to carry on this work, and the energy to follow through when it becomes overwhelming. Leaving Nairobi, I reminded myself what I always tell audiences: Hope is a moral imperative necessary for our work and commitment. That I had rediscovered such hope in the slums of Kibera and the rose plantations of Lake Naivasha simply deepened the mystery of it all for me.

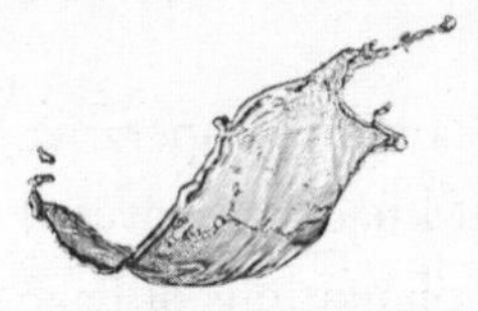

THE UNMENTIONABLES

Rose George

Rose George holds degrees in modern languages from the University of Oxford and international politics from the University of Pennsylvania, where she was a Fulbright scholar. She has written for the *Guardian, Independent, London Review of Books,* and *New York Times.* Her first book, *A Life Removed: Hunting for Refuge in the Modern World,* explored the reality of life on the run for Liberia's exiles; it was long-listed for the Lettre Ulysses Award for the Art of Reportage. Her latest book, *The Big Necessity: The Unmentionable World of Human Waste and Why It Matters,* examines the current state of sanitation in the world and makes a case for pushing poor sanitation—which kills more children under five than AIDS or malaria—up the global political agenda. *The Big Necessity* was judged by the *Economist* to be the best science book of 2008; the American Library Association voted it one of its books of the year.

I NEEDED THE BATHROOM. I assumed there was one, though I was at a spartan restaurant in the Ivory Coast, in a small town filled with refugees from next-door Liberia, where water came in buckets and you could buy towels secondhand. The waiter, a young Liberian man, only nodded when I asked. He took me off into the darkness to a one-room building, switched on the light, and left. There was a white-tiled floor, white-tiled walls, and that was all. No toilet, no hole, no clue. I went back outside to find him again and asked if he'd sent me to the right place. He smiled with sarcasm. Refugees don't have much fun, but he was having some now. "Do it on the floor. What do you expect? This isn't

America!" I felt foolish, embarrassed. I said that I was happy to use the bushes. I wasn't fussy. But he had already gone, laughing, into the darkness.

I needed the bathroom. I left the reading room of the British Library in central London and found a "ladies" a few yards away. If I had preferred, there was another one on the far side of the same floor, and more on the other five floors. By 6 p.m., after thousands of people had entered and exited the library and the toilets, the stalls were still clean. The doors still locked. There was warm water in the clean washbasins. I did what I had to do, then flushed the toilet and forgot it, immediately, because I could, and because all my life I have done no different.

That is why the Liberian waiter laughed at me. He thought that I thought a toilet was my right, when he knew it was a privilege.

It must be, when 2.6 billion people don't have sanitation. I don't mean that they have no toilet in their house and must use a public one with queues and fees. Or that they have an outhouse, or a rickety shack that empties into a filthy drain or pigsty. All that counts as sanitation, though not a safe variety. The people who have those are the fortunate ones. Four in ten people have no access to any latrine, toilet, bucket, or box. Instead, they defecate by train tracks and in forests. They do it in plastic bags and fling them through the air in narrow slum alleyways. If they are women, they get up at 4 a.m. to be able to do their business under cover of darkness for reasons of modesty, risking rape and snakebites. Four in ten people live in situations where they are surrounded by human excrement because it is in the bushes outside the village or in their city yards, left by children outside the back door. It is tramped back in on their feet, carried on fingers onto clothes, food, and drinking water.

The disease toll of this is stunning. A gram (0.04 ounce) of feces can contain 10 million viruses, 1 million bacteria, 1,000 parasite cysts, and 100 worm eggs. Bacteria can be beneficial: The human body needs bacteria to function, and only 10 percent of the cells in our body are actually human. But plenty are malign. Small fecal particles can contaminate water, food, cutlery, and shoes and be ingested, drunk, or eaten unwittingly. One sanitation specialist has estimated that people who live in areas with inadequate sanitation ingest 10 grams (0.4 ounce) of fecal matter every day. Poor sanitation, bad hygiene, and unsafe water—usually unsafe because it has fecal particles in it—cause one in ten of the world's illnesses. Children suffer most. Diarrhea—nearly 90 percent of which is caused by fecally contaminated food or water—kills a child every 15 seconds. The number of children who have died from diarrhea in the past decade exceeds the total number of people killed by armed conflict since the Second World War. Diarrhea, says the United Nations' children's agency, UNICEF, is the largest hurdle a small child in a developing country has to overcome, larger than AIDS, or TB, or malaria. An affliction that to most Westerners is the result of bad take-out food kills 2.2 million people annually—mostly children. Public health professionals talk about water-related diseases, but that is a euphemism for the truth. These are shit-related diseases.

You get used to euphemisms when you decide to write about sanitation. (There's one, for a start; "sanitation" can mean anything from garbage collection to sewer cleaning.) Sex can be talked about, probably because it usually requires company. Death has once again become conversational—so much so that it is given starring roles in smart, prime-time TV dramas. Yet defecation remains closed behind the words, all chosen for their clean association, that we now use to keep the most animal aspect of our body in the backyards of our discourse, where modernity has decided it belongs. Water closet. Bathroom. Restroom. Lavatory.

This conversational handicap has serious consequences. Politicians' unwillingness to talk about sanitation—and people's unwillingness to demand change—must influence the dismal current situation. Yet in the words of Umesh Pandey, a Nepali sanitation activist, "Just as HIV/AIDS cannot be discussed without talking frankly about sex, so the problem of sanitation cannot be discussed without talking frankly about shit."

In 2007, readers of the *British Medical Journal* were asked to vote for the biggest medical milestone of the past 200 years. Their choice was wide: antibiotics, penicillin, anesthesia, the pill. They chose sanitation. In poorly sewered 19th-century London, one child out of two died early. After toilets, sewers, and hand washing with soap became normal, child mortality dropped by a fifth. It was the largest reduction in child mortality in British history. In the poor world, proper disposal of human excreta—the process that is given the modern euphemism of "sanitation"—can reduce diarrhea by nearly 40 percent. Though 90 percent of most sanitation-related budgets goes to water supply, providing more or cleaner water reduces diarrhea by only 16 to 20 percent. Harvard University geneticist Gary Ruvkun believes the toilet is the single biggest variable in increasing human life span. Modern sanitation has added 20 years to the average human life. Good sanitation is also economically sensible; a government that provides adequate sanitation saves money on hospital visits avoided and does not lose labor days to dysentery or workers to cholera. Where good sanitation exists, people are wealthier, healthier, and cleaner.

It doesn't take an academic to work this out; a 60-year-old illiterate widow knows it perfectly well. A 60-year-old widow such as Sandhya Barui, one of the countless sanitation foot soldiers whom I met while writing my book, who made the writing of it easy. Sandhya lives in a small shack in a peri-urban slum an hour from Kolkata (formerly called Calcutta), India. Peri-urban means the slum looks quite nice, actually. There is space, and there are banana trees. Behind Sandhya's shack, there is a field of tall corn, which, on the hot day that we visit, looks cool and appealing. "That was my toilet," Sandhya told me. The leaves were tall, but never tall enough to hide her from laughing children. "They would come and look at you," she says, "and laugh while you squatted. It was shameful." In this daily humiliation, Sandhya was not remotely alone. Of India's one billion people, 750 million practice what is known as open defecation. This means they use whatever land or roadside is available. Driving through Indian villages—or even towns and cities—at dusk, there is a squatting safari. There: an old man squatting, his buttocks exposed. Over there: a woman who has stood up suddenly at the sight of a car, and hastily pulls down her sari, but whose face is resigned, not embarrassed. Here: two men walking back companionably, each holding a plastic lota that contained the water they have used to clean themselves.

Every day, 200,000 tons of human feces are deposited in India. I don't mean that they are dealt with, or sent down sewers, or given any treatment or containment. These 155,000 truckloads are left in the open to be trod on, stepped over, lived among. Open defecation in India is done on a scale, in the words of the sanitation NGO Sulabh International, that compares to "the entire European population sitting on their haunches from the Elbe in the east to the Pyrenees in the west." Indians sit on their haunches from the deepest forest to the heart of the cities. They do it beside train tracks, as novelist and essayist V. S. Naipaul recorded with disdain in 1964, and still do. The Indian journalist Chander Suta

Dogra recently described an early morning scene familiar to any train traveler in India: "Right in your face are scores of bare bottoms doing what they must." Open defecation is so endemic, people do it even outside public toilets in the center of a modern city. I see this one afternoon in the bustling city of Ahmadabad, where by early afternoon the pavement outside the toilet entrance is dotted with shit. I choose not to examine the interior.

An Indian businessman called Milon Nag tells me that when foreign clients came to visit his plastics factory near Pune, he kept a folder in the car, ready to hold up to distract his visitors from the sight of a roadside defecator. Nag is a considerate man, and the sight of people defecating in the open distressed him enough that he decided to do something about it. His accountant and sales manager became self-taught sanitation specialists. After years of research, his factory developed the most lightweight plastic latrine squatting platform on the market, now used by many aid agencies in emergencies. "For an Indian," Nag says, "open defecation is the maximum embarrassment." That's not quite true. The maximum embarrassment for Indians is trying to defecate in traffic or a wheat field, keeping one hand free to cleanse and yet managing to keep faces covered with saris, all the while watching for lurkers and lookers. Open defecation damages women most, because modesty requires them to do it under cover of darkness, leaving them vulnerable to sexual assault, snakes, disease, and infection. Blocking natural bodily functions like urination and defecation can cause bladder and urinary tract infections and worse.

In Sandhya's family, someone had diarrhea at least once a month. She was spending a hundred rupees (two dollars) a month on medicine, which was more than she could spare. "To us, even if we lose five paise [less than a tenth of a cent], that's a lot of money."

Then some visitors came to the slum. They weren't the first, and wouldn't be the last. Countless people had come to walk up and down the slum's unpaved streets. Sometimes they came under cover of darkness because there was no electricity, this being an illegally squatted patch of land, occupied by refugees from West Bengal's devastating floods 30 years earlier. The slum dwellers had done what they usually did, which was show their guests the nicest parts of their homes, and then the visitors left. But the visitors who changed Sandhya's life were different. They didn't just want to see the good bits. At the end of the village tour, they said, "Show us where you defecate." The villagers were shocked. But they took the guests to the outlying areas and bushes where they would do their business, day in and day out. The visitors nodded and walked back to the meeting space. They told the villagers to do some math. If the average human being defecated 250 grams (8.8 ounces) of feces a day, how much was being left around? The villagers may have been illiterate, but they were not stupid. They calculated that it was truckloads. Then they noticed that, sneakily, one of the visitors had set down a plate of chapatis (flat bread) next to some feces taken from the open defecation ground. And they noticed the flies flitting from one to the other, then landing on the villagers. Though they had been practicing open defecation for years—despite having enough land to build latrines—they had to face a shocking revelation. If the flies were landing on shit, then food, then the villagers were eating shit. In fact, the pioneer of this "triggering" method—known as Community-Led Total Sanitation, or CLTS—calculates that anyone who lives in a village that practices open defecation is ingesting 10 grams (0.4 ounce) of fecal matter every day, and that if even a single villager continued to defecate in the open, that would continue.

Sandhya wasn't the only one to be shocked. But she was the first one to set to work. Though she's slight, she's wiry, and stronger than she looks. Very shortly after that fateful village meeting, Sandhya was digging her own latrine pit. She paid 700 rupees ($15) for a latrine pan, because she had understood its value far exceeded its worth. She realized why her children kept being felled by diarrhea, and why even when she was feeding them well, they could never gain weight, with the worms and parasites washing the nutrients straight out. For the superstructure, Sandhya used banana leaves and cheap plastic. It's nothing much, her latrine, but it has been so revolutionary that she poses next to it for the camera, beaming. She doesn't have to be watched by children anymore while she attempts to conserve her modesty with her sari while also trying to clean herself with one hand (it's not an easy maneuver). She doesn't get diarrhea any more. And she has pride, and not just in her latrine. The rest of the village followed suit by installing latrines. People who continued to defecate in the open were shamed by children hounding them with whistles. In other villages the career of "toilet spy" was created, as denouncers of transgressors earned half of the 51-rupee (one-dollar) fine that was levied on open defecators.

Once the latrines were installed and the open defecation grounds cleaned up, the slum's youngsters set to work elsewhere. Drains were cleaned. Streets were kept swept. The local mayor, who couldn't acknowledge their illegal occupation of the land by giving them electricity or running water, gave them three solar streetlamps.

Sandhya is still not alone in her situation, and that's a good thing. India has some of the worst sanitation statistics in the world, but it also has some of the best activists, campaigners, and courageous women like Sandhya. India's Total Sanitation movement now has a national prize—the Nirmal Gram, or Clean Village—presented on national TV and handed out by the president of India. Such high-profile governmental interest is rare. Money generally goes to more fashionable diseases and causes, even though poor sanitation is linked to 25 percent of child deaths. In Madagascar, for example, 0.1 percent of the population has HIV/AIDS, and UNAIDS found there were too few AIDS deaths to estimate. Yet HIV/AIDS receives five times more funding than sanitation, though diarrhea kills 14,000 Madagascan children every year.

Sandhya, though, now knows something that it has taken health economists years to calculate: Investing in sanitation is a bargain. In fact, it's probably the best public health bargain there is. Investing one dollar in sanitation reaps eight dollars in health costs averted and labor days saved. Last year the World Bank calculated that poor sanitation cost Cambodia, Indonesia, the Philippines, and Vietnam 1.4 to 7.2 percent of their GDP. When Peru had a cholera outbreak in 1991, losses from tourism and agricultural revenue were three times greater than the total money spent on sanitation in the previous decade. Antiretroviral therapies, the most common prevention tool against AIDS, cost $922 per DALY (disability-adjusted life year, a standard calculation of disease burden). Sanitation and hygiene promotion cost $11 and $3, respectively.

In other Indian villages, women told me that once latrines and running water had been installed, 80 percent more girls were going to school. Sanitation improves school attendance, health, and well-being. It frees women to do income-generating work, as they're not fetching water or getting up before dawn. What Sandhya and thousands of other Indians who are improving their sanitation situation after CLTS and similar initiatives taught me is that sanitation isn't brought by development but can trigger it. If donors and politicians had a fraction of Sandhya's guts, the world's biggest unresolved public health crisis would be a lot closer to resolution.

BRIDES OF THE WELL

Shekhar Kapur

Shekhar Kapur is an internationally acclaimed Indian film director and producer. He began his career as an accountant and management consultant in Great Britain but soon abandoned his profession to pursue his artistic interests. His directorial debut came in the early 1980s with *Massoom* and later *Mr. India.* However it was his real-life story *Bandit Queen,* about a low-caste female outlaw who achieved notoriety, that brought him international acclaim and attention. Kapur was subsequently invited by members of the Hollywood film industry to direct *Elizabeth,* which was nominated for seven Oscars. He went on to direct Heath Ledger in an adaptation of the Edwardian novel *The Four Feathers* in 2002, and in 2007 directed *Elizabeth: The Golden Age,* which was nominated for two Oscars. After ten years in Hollywood, Kapur has now returned home to Mumbai, where he is pursuing his dream project *Paani (Water),* a movie about a futuristic city of 20 million people with no water.

ON THIS MORNING SARASWATI struggled to get out of bed. Well, I say morning because the birds had begun their morning melodies long before the beautiful hues of blue streaked the sky across the dry land. It was the one time that the land felt magical and mystical, a land claimed over the years by the flowing desert. Few shrubs remained to tell the story of long-gone days of the changing of the seasons from shades of green to golden and then brown. Furrows cut and burned into the white-caked mud told the tale of a river that once must have flowed.

No one spoke much in the village of Barmer. What needed to be done was simple and ritualistic. Nor was there the usual merriment of festive occasions provoked usually by the chaotic mating rituals of young men and women. For if heard carefully this was a village of older people. The dominant sounds of the day would be the dry, cracked sounds of older vocal cords, not contradicted by the clear singsong lyricism of young men and women, nor the bubbling sounds of little children.

Quietly Saraswati put on the bells of her anklets, making sure that the sound would not wake up her husband. She loved this sound, and she would walk with a harder step than normal, so that the other women at the well would be envious of her anklet bells. It was the only thing her parents could give when she left her village in a time that seemed so far, far away now. And as Saraswati walked by her husband, she rebelliously put her foot down hard to play with destiny a little. But she knew that the emaciated body ravaged by the desert and by age, snoring through an open, toothless mouth, would not wake up till the flies flowed uncomfortably across his mouth looking for leftovers.

But a thrill passed her every time she did that. Imagine if he woke up to discover that his wife was not at all the woman that he stored away in one corner of his mind! Saraswati rushed out, bent over, and coughed. Her back ached, but there was no escape. They said that this was how it was, but Saraswati at 15 had a heart that played a song with what lay on the horizon.

And so it was every day. Just as the hues of blue showed the silhouette of the village, young girls emerged like ghostly shadows from a fairy tale. These were the "child brides of the well," as they had become known to villages far away. The rivers and the wells retreated into their distant sanctuaries, leaving in their wake villages and communities desolate but for older people unable to move to the cities where they would have to fight for another existence, another life. But the caste system provided the solution, as it had done over the centuries, manipulating its own malleable, slithery body like an indestructible amoeba. So the clear oppressive lines of separation between people of various castes could become blurred but never actually disappear.

This was, after all, a village of higher castes.

The priests let it be known that for a young virgin to be married into a higher caste would absolve her whole family and their entire lineage of bondage into servitude. Young low-caste girls were consecrated by the priests in temples, and amid much ceremony, a procession of 20 young girls was sent from outlying villages to Barmer. It was a strange sight—young girls, nervous and giggly, walking into a village to be welcomed by bent old men and women, and anxiously looking for the high-caste young men who had agreed to marry them. Only when the marriage rites began, and as the drums played and the girls emerged from the huts with coy smiles on their faces, did they realize that the bent old bodies in tattered turbans were about to become their husbands.

Saraswati remembered Paras, who ran away screaming half naked the next day. She was just 12. Three weeks later she returned. Her family had closed the doors to her, busy as they were paying obeisance to the higher gods of the high-caste community. She was sent to the temple to be purified by the priest of her sins, the rituals of which had gone on for three nightmarish weeks. Finally Paras had nowhere to go but to the place she was told that her new gods and her karma had deigned for her—the village of Barmer, to her 73-year-old husband.

Saraswati and most of the other girls were more fortunate. Their husbands had little interest in their young bodies, or the energy to indulge, even if they did. But there were more immediately pressing needs. Some of the old people needed care even in their daily chores. The houses needed to be cleaned, and meager kitchens needed to be kept going. But beyond that there was a more fundamental need that the girls had been brought for: water.

The nearest working well was nine miles away. There was no path, even, and the only way to get there was on foot. That's how they'd come to be called the child brides of the well. Each day they walked four hours to the well in time to meet the scorching heat of the midday sun, and back five hours laden with pitchers of water—often balanced precariously—through shifting and uncertain sand dunes that had begun to invade from the deserts in western wastelands.

And so they would walk, for the rest of their young lives.

But there was something about Saraswati this morning. Paras was intrigued. For three years they had walked together to the well, apart from the other girls and mostly in silence. After all, there was not much that could provide fodder for gossip for young girls in Barmer, and very little drama, except for when Paras's mother-in-law had started to beat her in a drunken state. In a fit of rage Paras had slapped her back, and the whole village decided that she needed to be taught a lesson. For two whole days Paras was not allowed a single drop of water. Her innate rebelliousness had to be quelled.

Then there was the time when Paras and Saraswati had used a couple of handfuls of water to drink and wash themselves just before they entered the village. That was a secret that bonded them—that they had used the water they'd carried all the way back to the village for their own use, before it had been handed over to the elders.

As the evening shadows came, Paras and Saraswati would approach the village after hauling their now full pots of water. Exhausted, they would pause by the lone tree at the outskirts of the village and pray fervently. When spotted they would swear they were dutiful wives praying for the long lives of their creaking husbands. But the prayers were secretly directed toward a different god. Rather than the god of eternal youth, they would be praying to the god of water, praying for the well to dry up.

The gods seemed to be answering their prayers; the well was going dry. The next well was too far to comprehend. When the water ran out, the girls would be freed. Their village would finally die out and the young girls, no longer needed, would be free to go. Having fulfilled their karma, they would be delivered by the high-caste gods to a different destiny.

The well was used by all 14 villages in the three districts that it served. Only one of those villages would get water supplied directly in a tanker pulled by two tired cows, as a dirt track still led to that village. That was because the distant cousin of the mistress of a local politician owned land there and would visit with friends in a noisy modern four-wheeler. That was always an occasion, because the villagers would pick up empty beer bottles left in their wake. Anything to store water in. Paras and Saraswati often wondered at the exciting lives of the young girls who went to that village to get married.

Paras still wondered at the spring in Saraswati's step today. The bells on her anklets seemed to beckon even the birds to gossip. Paras wanted to know what secrets the birds shared with Saraswati. Her footsteps on the parched earth were no longer the rhythm of a plodding cow. The parched earth seemed to come alive with the uncertain dance of each step.

But Saraswati would not tell. She just giggled and put distance between her and Paras. The shadows cast by the early sun were still long enough to connect the two, and Paras tried to capture the secret by constantly tugging at Saraswati's shadow. But then Saraswati took her pitcher off her head and lay down on her back, stretching her arms wide to feel the coolness of the yet young earth on her body. The shadow was gone, and Paras suddenly felt completely naked. Never before had she taken this journey without the comfort of another shadow always walking by her side. The rhythm that kept them going these years was suddenly broken.

Now, if you were a vulture swooping down to investigate, you would be forgiven for being confused. For lying perfectly still, hands stretched in the vast, flat yellow landscape, were two young bodies. It is not often you saw potential food so still, yet breathing life as if they had just discovered it.

Paras felt as if she could hear Saraswati's wild heart beat through the fluid earth. She felt hers responding, afraid that miles away, back in the village they would hear their rebellion again.

"You did?" Paras was almost afraid of the next word: "When?"

Their fingers touched. Lost in some imaginative world, Saraswati gently ran her tongue across her now dry lips, as if parts of her body were still dancing to the intensity of those moments. A bead of sweat trickling down her forehead, Saraswati gently led Paras's hand to her heart and laid it there so she could hear the truth through its wild throbbing.

"And . . . ?"

Saraswati squeezed Paras's hand so hard that Paras cried out in pain, panicked, and tried to escape. But Saraswati suddenly leaned over and looked straight into Paras's eyes, holding them with a fierceness and intensity that told the story of the unimaginable.

Something changed that moment. Did the winds pick up, carrying Saraswati's words across the land to her lover? The birds went wild, confused at centuries-old rules being broken. The desert resonated with Saraswati's breathless words as she poured out every acute memory of her encounter, with absolute intimacy.

Not even the gods, or centuries-old tradition, had the power to stop a young girl's discovery of her feminine self.

"Who—" Caught in the first flush of Saraswati's forbidden words, Paras was now panicking.

"The boy . . . ," Saraswati began, suddenly coy. Had she revealed too much? Would Paras possibly carry the secret in her belly forever? But Saraswati was feeling brave today. She felt a surge of power. "The boy that comes every six months with his father to sell medicinal oils."

It was all too real for Paras now. The panic swept up, engulfing. She leaped up and screamed at Saraswati: "Sin! Sin!" The vulture squawked as the birds died down. Paras kicked dirt into Saraswati's face, again and again. The sun was stronger. Higher. The shadows were much, much shorter.

Saraswati ran after Paras, the pitcher balanced precariously on her head, desperately trying to keep up with Paras's shadow. For where could she go without it? "I will die if you tell," Saraswati screamed. "I will deny it! The whole village will know you are a liar."

The wind was not listening anymore. The birds had lost interest. The vulture looked for other prey. The sun, directly overhead now, was casting no shadow. Paras and Saraswati were free of each other, but Saraswati kept shouting, till she was hoarser than the morning crows.

Paras whirled around and slapped Saraswati hard, so hard that Saraswati's pitcher fell down. But even then the instinctive laws of water kicked in. Paras caught the pitcher in time and roughly handed it back to Saraswati.

"He swore—he swore—I was the only one." Paras's confession was not as passionate as Saraswati's but just as fierce.

The shadows were long again as the sun wilted and got tired of the hot day. But Saraswati and Paras no longer cared to be in each other's shadow as the well came into sight. Nor did they pay much attention to the 50-odd women fighting for the narrow space on the perimeter of the well. The pitchers defined the caste of the women. The upper-caste ones had stronger brass pitchers, but even though adopted into the higher caste, Saraswati and Paras could only afford clay pitchers. It was a struggle to get your pitcher into the well and yet avoid having it smashed against the brass ones or the side of the well. If a dispute broke out, high-caste women always had first right. It was tradition so ingrained that the thought of confrontation never even occurred to Paras or Saraswati as they endured the constant abuse and curses. But this was a daily chore, and both girls went through the paces, with other matters on their minds.

Paras carefully watched her footsteps. She carried a much heavier load on her head than when she started. Balancing her pitcher, she wondered if Saraswati still had a spring in her step. She had left Saraswati far enough behind for her not to notice. Paras tried a spring in her own step, like a little dance. The full pitcher almost fell and Paras caught it just in time. But a little laugh escaped her.

"Paras!!"

Paras froze. Had Saraswati noticed her dance? She looked around, and, pitcher carefully balanced on her head, Saraswati was running toward her. Secretly Paras was glad. Five hours was a lonely walk back without another shadow to keep you company.

Saraswati came up to Paras. She looked down and danced a little step, daring Paras to do the same. Paras did, and the two young girls, having discovered a common spring in their step, giggled.

"He's not coming back for six months," said Paras.

"And we will be on this journey every day," replied Saraswati.

"For the rest of our lives," Paras said sadly.

"No," replied the now optimistic Saraswati. "Only till the well runs dry."

"Only till the well runs dry," agreed Paras, as both the girls lowered their pitchers and knelt in fervent prayers.

The village of Barmer was creaking to a halt, getting ready to give up on the rigors of the day, hoping the dreams of the night would provide an escape to those who could sleep. They searched anxiously for the last two girls to return from the well, needing the water and their young hands to do the nightly chores. In the distance the villagers saw one long shadow. Just one.

Had one of the girls run away? Moans and tired curses escaped the lips of those who imagined the chores that would be left. Already there was talk of how to make one girl do the work of two.

But those who looked carefully would have seen two girls, their hands on each other's shoulders. A spring in their step.

Two girls and one shadow.

A WORLD THAT WORKS

Bill McKibben

Bill McKibben is the author of a dozen books about the environment, including *The End of Nature,* regarded as the first book about global warming for a general audience. His work has appeared frequently in publications such as the *New Yorker,* the *New York Review of Books, Orion, Harpers,* the *Atlantic, National Geographic,* and *Mother Jones.* A scholar in residence at Middlebury College, McKibben has received honorary degrees from a wide range of American colleges and has been the recipient of Guggenheim and Lyndhurst Fellowships. He has spent the past few years coordinating the biggest national and international campaigns on global warming.

IT'S EARLY MORNING IN VARANASI. Thousands of residents of this crowded city have come down to the Ganges to bathe, lathering up and rinsing off, or pounding their laundry against rocks along the shore. Thousands more pilgrims are arriving, as they do every day, streaming in from across India to immerse themselves in the sacred waters of this river. Wood for funeral pyres is piled by the bank; smoke drifts out of the temples where cremation has been going on all night. Meanwhile, the newspaper seller is shouting the day's headline: Drought has been declared across this part of Uttar Pradesh, as the monsoon rains have failed to appear.

If there's a place on this planet that makes clear how central, how complex, and how personal the water crisis is, this may be it. For the scene is almost literally timeless—this is one of the oldest pilgrimage sites on Earth, and the tableau is largely unchanged from what you would have seen 300 or 400 years ago, and probably ten times further back than that. I'm here on a pilgrimage of my own, helping build the global grassroots movement against climate change—but today change of any kind seems far away. The sun rises orange against a hazy summer sky; the riverbank is filled not only with people but with cows, goats, donkeys. The ancient cycles of birth and death roll on—when we pull our boat to the opposite bank, we come ashore next to a human skull floating on the gentle ripples.

But if the scene is in some sense out of another age, it's also completely tangled up in our problematic present. In fact, unless we make some very serious changes very soon, the odds are slim that this scene will exist much longer. A confluence of woes all piling up at the same moment threatens to wreck this river—and end both the practical and symbolic roles it plays in the lives of so many. The various dilemmas can't be thought of separately—they layer on top of each other, a kind of geological profile of the way the water crisis has mounted slowly and almost invisibly over the years. Now action can't be put off any longer.

The simplest and most obvious crisis—the one that defines how we used to think about water troubles—stems from the fact that Varanasi (and indeed most of the other cities along the Ganges) doesn't effectively treat its sewage before it enters the river. Either there's no plant at all, or it shuts off during the incessant power outages or is overwhelmed by the rain when the monsoons actually appear. The result is that the river is filthy, with bacteria counts so far off the charts that people are routinely sickened. (Pilgrims aren't supposed merely to immerse themselves for a holy "cleansing," but to take a sip of the polluted water.) This state of affairs has persisted for decades, despite big expenditures by the Indian government, money that seems mostly to have been wasted or stolen. It's not that different from American rivers and streams a generation ago—although at least the Ganges has yet to catch on fire. But a lot of laws and a lot of money have helped to clean the worst of North America's pollution.

The same could happen here, of course—and if you spend much time with the folks I've been talking to, you start to feel it won't be long. The Sankat Mochan Foundation is one of the most interesting environmental groups I've ever come across. Its leader, V. B. Mishra, is both a Hindu spiritual leader and a water engineer. Mahantji, as he's known, has gathered around him a cadre of mostly retired scientists from nearby Banares Hindu University, and together they've mounted a quarter-century campaign for a cleaner river. They've agitated for a particular solution: a lower tech, lower cost, and more natural series of settling ponds that should produce clean water. In the summer of 2008 the Indian government finally agreed to fund a demonstration project. And so the mood inside their headquarters, just up one of the ghats, or stone steps, from the riverbank, is ebullient. "It is a miracle," says Mahantji, although he quickly adds that it's an unfinished miracle—now they have to prove that their solution will actually work better than the millions of rupees already wasted on sewage treatment. It's the kind of miracle that comes when you work really, really hard for a really long time and when you're able to involve masses of people who may not know much about water chemistry but do know that the Ganges is the mother river of their faith. It makes sense that their headquarters is a temple.

Hard as that fight has been, however, it almost certainly won't be enough to solve the problems now attacking the river and its neighbors. India in general and Uttar Pradesh in particular have growing water shortages—the aquifers have been overpumped for irrigation, and climate change seems to be starting to play havoc with the vital monsoons. They arrive late, if at all, and the rain comes spottily, sometimes in unprecedented deluges that hurt as much as they help. Right now the water level is pitifully low, and crops are withering in the fields. It's not hard to convince anyone here about climate change. They may not know that warm air holds more water vapor than cold, and hence evaporation is on the rise and with it drought, or that eventually all that water will fall somewhere, increasingly in the kinds of torrents that wash away everything in their path. But they do know that the farming their ancestors have done for a very long time is increasingly difficult, even impossible.

And that's before the biggest trouble of all kicks in. The source of the Ganges is Gangotri, a huge glacier upstream in the Himalaya. In Hindu mythology, it's where the goddess Ganga came to Earth from the heavens, in the form of a river; Lord Shiva caught the water in his matted hair to absorb the force of its fall. It's a singular place, in other words, but just like every high-altitude glacier on Earth, it's melting rapidly as the climate warms.

That glacial melt is one of the most daunting effects of rapid climate change. For the more than 10,000 years of the Holocene, it has been warm enough to keep the continental interiors free of ice, and hence open for agriculture. But it has been cold enough that the Andes and the Alps and the Sierra have kept frozen reservoirs of ice, which slowly melts each summer, producing the river flow that provides irrigation and drinking water in the hottest months. Nowhere is this clearer than in the Himalaya, where a compact area of glaciers not much larger than Italy provides the headwaters for the Salween, the Mekong, the Brahmaputra, the Yellow, the Yangtze—and the Ganges. One human being in three lives downstream—and that's *why* they live there. Without those rivers, life would be impossible.

Gangotri may melt away by mid-century, by some estimates. The Indian government disputes this, insisting that although the ice is melting, it's too early to blame it on climate change, and predicting the thaw will slow. But the experience around the world is not encouraging—in 2009, for instance, scientists looked on as Bolivia's highest glacier, Chacaltaya, melted away entirely. In June, glaciologists held a special ceremony to mark its passing; meanwhile, down below, city officials were scrambling to figure out how to replace the water that's suddenly gone missing. If Gangotri disappears, the Ganges won't dry up entirely, but it may become seasonal, and its flow will be greatly reduced. Which, of course, will make every other problem it faces that much worse—the same amount of sewage in half the quantity of water stinks twice as bad. If the rains aren't falling *and* the river turns sluggish, farmers are even more out of luck.

You could tell this same story the world around, of course—you could tell it about the Murray-Darling River system in Australia, or the Colorado in the American Southwest. (Some recent predictions foresee Lake Mead, which backs up behind the Hoover Dam, as having a 50 percent chance of drying up altogether by 2021.) But at least the good people of Sydney and Las Vegas did it to themselves—it's their carbon up there in the atmosphere that's raising the temperature and turning off the tap. Varanasi? Not really. Yes, the streets of the city are choked with traffic—it takes a very long time to get anywhere. But most of the traffic is bicycle; a certain amount is horse cart; and then there are an enormous number of people walking—barefoot, many of them, in the orange outfit of the pilgrim or the sage.

In some very real sense, the richest people of the world are stealing water from some of the poorest. Not just not helping them—*hurting* them. Taking places that have worked for eons and breaking them.

Our politics will be filled with battles about climate change in the years to come, and one of the refrains we'll hear from demagogues is: The West shouldn't do anything to rein in its emissions unless China and India do just as much. When I hear that refrain, I'll think of the scene along the ghats on this stretch of this Ganges. I'll think of Mahantji and his crew of septuagenarian sewer engineers gamely figuring out the next step in their battle to save the river. I'll think of a glacier that's been there since the Hindu world began. I'll think of a world that's worked for a long time without our even worrying about it.

PHOTOGRAPHY JOURNEY

Alton C. Byers

Alton C. Byers is a mountain geographer, climber, and photographer specializing in high-altitude conservation programs, applied research, and climate change impacts in the mountains. He received his doctorate from the University of Colorado in 1987, focusing on landscape change, soil erosion, and vegetation dynamics in the Sagarmatha (Mount Everest) National Park, Khumbu, Nepal. He has worked for the Mountain Institute since 1990 in its Asian, Andean, and North American programs and has published widely on a variety of field research topics. His recent awards include the Nature Conservancy's Mountain Stewardship Award; American Alpine Club's David Brower Conservation Award; Association of American Geographers' Distinguished Career Award; and Sir Edmund Hillary Mountain Legacy Medal for "distinguished service to mountain people and environments in remote regions." Since 2007 he has directed the Alpine Conservation Partnership project, which, in collaboration with the American Alpine Club, works to conserve and restore fragile alpine ecosystems worldwide.

I FIRST SAW THE MOUNT EVEREST region of Nepal as a 21-year-old in 1973. That year I spent several months trekking throughout what would soon become a national park, working on a research project that studied trekking and climbing impacts on Nepal's mountain environments. Since then, I have been fortunate to have a long and fruitful affiliation with that area's people and high mountain landscapes. It was there in the spring of 1980 that I met my future wife, Elizabeth, who was part of the trekking group I was co-leading. We returned as a married couple in 1984 and lived for a year in Khumjung village, a cluster of about 50 traditional Sherpa stone houses nestled beneath the slopes of the sacred mountain Khumbila, protector of the Everest region. There I conducted the field research for my doctorate, measuring soil erosion between the main Sherpa settlement of Namche Bazaar and the upper alpine zone near the Everest base camp. I also developed a keen interest in understanding how these mountain landscapes change over time: how vegetation thrives or suffers, how glaciers may advance or retreat, and how this affects the watershed and the fragile alpine ecosystem. I study this with a variety of techniques: carbon dating, soil analysis, oral testimony, and repeat photography—comparing images of the same regions taken years apart.

Interested as I am in historical photography of Mount Everest, I was handed a treasure when in 1999 my former adviser in graduate school, Jack D. Ives, gave me a box of old photographs. Taken around 1950, these photos of glaciers in the Mount Everest region of Nepal had been the property of the Swiss-Canadian glaciologist Fritz Müller. The box also included Müller's old field notes; pressed plants; rock rubbings; letters; telegrams from New Delhi, India, regarding travel arrangements; sketched mountain panoramas with the name of each summit carefully penciled in; and several dozen labeled manila envelopes. Müller had been the scientific team leader of the successful 1956 Swiss Everest expedition, and although not attached to the climbers, he was tremendously strong and ascended to altitudes of more than 26,900 feet, helping the Sherpa high-altitude porters carry loads to the South Col "on his day off." When the climbers went home, he stayed on for an additional nine months, living at altitudes above 16,400 feet, which at the time was a nonindigenous record. Müller suffered a fatal heart attack in 1980 at the age of 54 while on the Rhone glacier, protesting the construction of a new hydroelectric facility with two of his graduate students, and in the confusion that followed, all of his photographs of the Everest region were lost. All, that is, except a single box of photographs that was salvaged from Müller's home by Konrad Steffen, then a graduate student of Müller's, who brought the box to Boulder, Colorado, in the early 1980s.

As it turned out, the box of photographs was an incredible window into the past. One packet contained fascinating snapshots of Müller's Sherpa research assistants at work, coring ice, carrying loads to the high climber camps, or exploring different ice caves and formations. Others contained the landscape photographs of Austrian climber-cartographer Erwin Schneider, taken in the 1950s in the course of creating his famous map of the Everest region. Schneider's stunning high-mountain landscape panoramas, made of individual photographs meticulously taped together and perfectly folded like an accordion, were contained in dozens of manila envelopes.

The photographs were not only of immense historic value, but in some cases were the only remaining prints left in the world because of the loss of the original negatives and archives. They were also a unique tool for understanding how climate change might be affecting the area. But even only eight years ago, "climate change" was not in most people's vocabulary. It was still controversial, the subject of heated debates, and there was little funding available to launch any kind of climate change-related expedition. Even my own 1998 replication of Schneider's photographs of Peru's Cordillera Blanca (White Range) in 1932, showing dramatic glacial recession, generated little interest beyond the occasional comment of "Wow, the glaciers are melting."

But by 2007 and with the release of Al Gore's movie *An Inconvenient Truth,* climate change had become the topic of the day, and I was pleased when the American Alpine Club offered to help co-finance a Khumbu expedition in August 2007 as part of its growing concern about glacial recession. By piecing together a number of other pockets of funding from my other projects, I was able to cover most of the costs involved in what was to become a 60-day expedition in the fall of 2007 and spring of 2008. I would walk hundreds of miles retracing the footsteps of Müller, Schneider, and Evans, and seeing through their eyes what changes had occurred during the past 50 years. These two field expeditions, it turned out, would also provide an opportunity to informally interview dozens of Sherpa people encountered along the way about climate change, as well as to study the causes of, and influences on, the melting of glaciers and glacial ice that I was about to witness.

Things did not start well on this particular expedition. All flights to the Everest region in early October 2007 were canceled because of bad weather, and during what was supposed to have been the first day in the field I was stuck in the Kathmandu domestic-flight airport. However, I was finally able to fly out to the 9,400-foot-high Lukla airstrip, which Ed Hillary had built in the 1960s high on an ancient river terrace. But the weather remained bad for the next week as I trekked from the airstrip up to the higher altitudes on the trail to the Everest base camp region. Day after day the Himalayan peaks remained hidden in clouds, and on more than one occasion I couldn't help but lament, late at night, "What in the world am I going to have to show to the people supporting this work if I walk out of here with no photographs?"

Likewise, the porters were deadly slow, including one that I had hired the previous spring to trek back into the Hinku Khola in the Makalu-Barun National Park to conduct a survey of climber impacts on the fragile alpine ecosystems there. When I asked what the problem was, he said that he had hurt his foot the previous week, so I gave him only the laptop to carry. It quickly became apparent that I couldn't act as both field researcher and sirdar (trek leader) at the same time and get any work done. Upon arriving in Dingboche, I therefore asked Sonam Ishi Sherpa, research assistant to many a noted Himalayan scholar, for help in finding a sirdar type to take over the job of porter supervision, buying supplies, and arranging the day's agenda. He recommended Everest summiter Pema Temba Sherpa, who, as it turned out, was a fantastic sirdar, immediately taking charge and sorting out all logistics related to gear, food, and travel schedule.

Even then, problems continued to surface on an almost daily basis, largely related to my lack of experience with digital cameras and the new necessities of frequently downloading photos and charging batteries, in my laptop as well as my camera, in one of the more remote regions of the world. More than one sleepless night was spent camped somewhere below a high-altitude pass or in a campsite high in the mountains, wondering how in the world I was going to (a) recharge my laptop battery, (b) recharge the camera batteries, (c) download the photos onto a four-year-old computer with "extremely low memory" messages surfacing every few minutes, and (d) back up the photos.

I would start a typical day by rising around 6 a.m., eating a breakfast of porridge or rice, and departing by 7 a.m. with Pema. Pema and I normally reached the photopoints—prominent hilltops with good views, usually between 16,400 and 18,700 feet, sometimes higher—by 11 a.m. Finding the actual photopoints was something of an art. Having spent years in the Khumbu and other high mountain landscapes of the world, I can often look at an old photograph and know with some certainty where it was taken, although dozens in my collection still have me baffled. Sherpas are often very good at recognizing the landscapes and landmarks, probably from their days as children herding yaks or working for climbing expeditions, but even here there were times when Pema didn't have a clue where the photo was taken. Many times I was convinced that the photopoint was the summit of a particular mountain in front of us, only to reach the top after hours and hours of climbing, look at the old photo, look at the prospect before me, and say, "Uh-oh—wrong place." Only then would I realize that it was the *next* hill to the north, necessitating a descent of 3,000 feet or more and, once again, a climb up to the next possibility.

On several occasions, the envelopes holding the panoramas were mislabeled, sending us up the wrong valley on a wild-goose chase. But we still returned to base camp with a beautiful set of photographs that might make an interesting color panorama, even if it didn't replicate an older photograph.

Sometimes the photopoints, moderately challenging in good weather, became extremely dangerous to reach when new snow made the steep rock slopes extremely slick. At times like this one had no other choice than to settle for a nearby photopoint as being good enough for the objectives at hand, and to note this in the record. I had to remind myself that Müller, Schneider, and Evans had between six and eight months and six field expeditions, respectively, to do their work, and I had only a combined total of 60 days during two field seasons. Nevertheless, I felt really fortunate that I was able to cover and photograph as much as I did during the fall of 2007 and spring of 2008.

Sometimes Müller and Schneider built cairns, or piles of stones set up by mountaineers as landmarks, to mark the location of the photopoint. During at least one or two occasions I found an ancient-looking tin can or other artifacts that had to have been left by one or the other, probably Schneider. Pema Temba agreed that the finds were of comparatively ancient vintage, clearly from the 1950s. It made me think of my colleague Conrad Anker, the world-class alpinist who found the body of British climber George Mallory on the north side of Everest in 1999, a figure lost to the world since 1924 when Noel Odell last saw Mallory and young Andrew Irvine vanish in the clouds below the summit of Everest. Thankfully, I didn't encounter any dead climbers' bodies, but I may have found Erwin Schneider's lunch.

All of this climbing and trekking and camping throughout the Everest region gave me ample opportunity to get the perspectives of lodge owners, yak herders, and other Sherpas about climate change during the past 20 to 30 years. Although these were informal interviews at best, it was interesting to note the clusters of agreement that started to fall out from everyone I spoke to, which are similar to nearly every other region of the mountain world where I've interviewed people. They would say, for instance, that weather had become much more variable and unpredictable (you couldn't count on the rains' coming on the same date anymore); that there were more climatic extremes than before (drought for weeks, followed by torrential downpours and floods; no snow for weeks, followed by a blizzard that taxed the roofs of even the strongest homes; and extremely cold winters compared with previous years); and that agriculture and people's incomes were suffering as a result.

It was also clear that much had changed in the Khumbu during the past 50 years, especially the glaciers. For example, what had been the Imja glacier in the 1950s was now a lake, with the trough created by the original advancing glacier now filling up with millions of cubic meters of water as the glacier receded. The people we spoke to were alarmed by this rapidly expanding Imja glacial lake and by the possibility of a catastrophic glacial lake outburst flood. Many had experienced the destruction of homes, property, and infrastructure caused by the outbreak of the Langmoche glacial lake in 1985 and wanted nothing to do with the possibility of a new outburst.

In fact, during the past several decades in the Everest region, research has revealed that 24 new glacial lakes have formed and 34 major lakes have grown substantially as a result of climate change and regional warming trends. Twelve of the new lakes have been tentatively classified as "dangerous," based on their size and rapid growth over the past several decades, as determined by satellite image analysis.

I also found that dozens of small, clean glaciers had disappeared throughout the region. It was clear to me that these glaciers, lacking a protective insulating cover of debris, located at lower altitudes (below 19,700 feet), and without a source of fresh snow and ice, were the most vulnerable to the impacts of warming trends. The same situation can be found with the rapidly disappearing glaciers in Glacier National Park in the United States, Huascarán National Park in Peru, the European Alps, and the volcanoes of Bolivia and Ecuador. Thankfully, much of the ice above 19,700 feet in the Everest region, at least on north-facing slopes, appeared to remain intact and largely undisturbed for the time being, thanks to the cold annual temperatures at these extreme altitudes. Ice with south- or southwest-facing slopes did not appear to fare as well.

Typical of most other regions of the world, I found that no detailed, on-the-ground field studies of potential climate-change impacts on the people and environments in the Khumbu—on their water, agricultural systems, safety, glacial lakes, or tourism—existed at the time. There had been many studies comparing satellite images, but none using the "muddy boots" methods of the old-time field geographer. Thankfully, recognizing the diversity and complexity of mountain ecosystems and cultures, scientists are increasingly advocating for exactly this sort of approach throughout the mountain world.

During April and May 2008 I returned to the Khumbu to participate as the lead scientist and photographer in the development of the "world's highest climate change photo exhibit." At base camp, a collection of my comparative panoramas, Müller's photographs of his Sherpa research assistants, and some of my older work showed physical and cultural landscape change. The Sherpa climbers and high-altitude porters loved the exhibit and spent hours pointing out changes in snow and ice cover that had occurred during the course of their own climbing careers. Combining art and science, the exhibit conveys three important messages, whether displayed in the Everest base camp, Barcelona, Berne, or Kathmandu: one, that climate change is affecting even the highest mountains in the world; two, that the lowland and coastal regions where most people live could experience severe water shortages; and three, that we can minimize these and other impacts of climate change by taking better care of our mountains and the mountain watersheds that provide the world with most of its fresh water.

I don't recall anyone in the Everest region being worried about water supplies, probably because with the acceleration of glacial melt there's been an increase in water that will continue until the ice is diminished or gone. Downstream, however, governments and donors are becoming increasingly concerned about the implications of their rapidly receding glaciers and the possible impacts on future freshwater supplies for millions of rural and urban residents. The loss of this snow and ice over the next 50 years could mean changes in water supplies as we currently know them, in the way that people use water and, of greatest importance, in the way that we treat and manage our mountain watersheds.

We probably can't stop the glaciers from melting, so what *can* we do to help mountain and lowland people deal with climate change, and to ensure that we have enough water for future generations?

We can help mountain people become more resilient in the face of climate change. A recent Mountain Institute project in the Andes, for instance, is helping local people develop ecotourism along the Inca Trail. Governments in the Himalaya-Hindu Kush concerned about the dangers of new glacier lakes have much to learn from the Peruvians, who have been controlling dangerous lakes in the Andes for more than 40 years. Building climate change awareness can help people begin to understand the critical importance of mountains to their lives and livelihoods—especially in the densely populated lowland regions that depend almost entirely on the mountains for their very existence. Collectively, these actions could help to create a new advocacy for mountains among highland and lowland populations, focusing international and governmental attention on the conservation and protection of mountain ecosystems worldwide.

Mountain conservation doesn't have to cost a lot, and tangible results are not necessarily hard to achieve. For example, with support from the Mountain Institute and the American Alpine Club, my Sherpa colleagues and I established the Alpine Conservation Partnership project in 2003. This is the park's first locally led initiative to protect and restore the region's alpine ecosystems, so heavily damaged during the past 30 years by uncontrolled adventure tourism. So far it seems to be an excellent example of how a lot of passion, a little money, participatory approaches, and some hard work can reverse the trends of environmental degradation.

This initiative came about after I led a National Geographic–funded research expedition to the Everest region in 2001 to study the impact of adventure tourism on alpine ecosystems. The results were disturbing: During the previous 20 years, more than 50 percent of the region's shrub juniper and alpine cushion plants had been cut and pulled up for fuel by tourist lodges and mountaineering expeditions. Juniper was being cut at a rate of more than 176,000 pounds per year. Cords of stacked juniper could be seen outside every tourist lodge; thousands of pounds more were cut for *puja* (worship) ceremonies, which asked the gods for protection before the climb. Porters, left to fend for themselves at the end of the day, were sleeping in caves and burning the fragile alpine juniper and dwarf rhododendron for warmth. It takes more than 100 years for a juniper shrub to reach a diameter of only one and a half inches. Already much of the alpine landscape was looking like a high-altitude wasteland.

But after community consultations and small seed grants from the American Alpine Club and the Mountain Institute in 2003, things started to change. Under the leadership of TMI's senior program manager, Ang Rita Sherpa, the Khumbu Alpine Conservation Council (KACC) was established. This locally constituted council is the world's first alpine NGO committed to protecting and restoring high-altitude ecosystems. The KACC immediately stopped the harvesting of all shrub juniper. A kerosene and stove depot was established to provide alternative fuel for tourists and lodges. A porters' rest house was restored in Lobujya to provide shelter, warmth, and cooking facilities for porters. (When I last checked last spring, it was filled to capacity every night.) New high-altitude nurseries are beginning the process of restoring devegetated hill slopes, and a 2.5-acre, cattleproof demonstration enclosure at Dingboche is allowing alpine hill slopes to heal themselves. New conservation curricula for local schools have been developed, training in project management is given regularly, and new income-generating products—such as a bottled juice, made from the berries of the local sea buckthorn shrub, which the trekkers love for breakfast—are being introduced.

In the 1950s, Müller, Schneider, and Evans were probably more concerned about the return of a Little Ice Age, and the destruction of forests and villages from the brutal and relentless *advance* of glaciers, than they were about the remote possibility of a general global meltdown. But it was painfully clear from my replication of their photographs, interviews, literature reviews, and general observations in the course of two field expeditions that the latter was precisely what was happening, even in the world's highest mountains.

The actual long-term impacts of climate change are still largely unknown. But I strongly believe that we as a people and as a civilization will learn to adapt to these changes, one way or the other. I believe that we can help by promoting good science that in turn will lead to good management. We need to come up with practical and effective adaptive strategies and mechanisms. We should promote interdisciplinary approaches—that is, those that include both the physical and the social sciences and the participation of local people—to thoroughly understand climate change impacts and possible remedies.

Finally, the one thing that everyone can contribute to is the conservation, protection, and restoration of mountain ecosystems and watersheds. This might begin by building awareness in highland-lowland interactions, such as through the photo exhibits mentioned previously, since most people are still unaware of the critical roles that mountains play in their daily lives, livelihoods, and survival. Ultimately, improved governmental and public recognition; new highland-lowland linkages and policies; large-scale conservation programs; and new, creative means of funding will give us the tools to protect our water supplies and the other ecosystem services provided by the mountains in perpetuity.

And that will be a pretty good legacy.

WHAT'S IN A NAME

Melanie Stiassny

Melanie Stiassny grew up in the south of England and received her B.S. and Ph.D. at the University of London. After serving on the faculty at Harvard University, she relocated to New York, where she is the Axelrod Research Curator (Ichthyology) and a professor at the Richard Gilder Graduate School at the American Museum of Natural History. Stiassny's research on fish biology and evolution has taken her throughout the world's tropical waters, but somehow she always returns to Africa, where her professional career began. Her current focus is on the Congo River Basin, where she has active field and education programs. She serves as an adviser to numerous scientific and conservation organizations, including the National Geographic Society's Conservation Trust, the World Wildlife Fund, Conservation International, and the World Resources Institute. One of her proudest moments was when an order of fish was named in her honor—the Stiassnyformes.

I AM REALLY NOT AT ALL SURE how my relationship with water began, but I'm guessing that perhaps it had something to do with my very first pet—a small fish collected in a local pond in the south of England, where I grew up. Back home such small freshwater fish are called tiddlers, and although the name strictly refers to the ubiquitous three-spined stickleback (*Gasterosteus aculeatus* Linnaeus, 1758), it is also used to refer to any small critter collected by kids in neighborhood ponds and streams. Going "tiddling" is one of the cherished memories of my early childhood years. Of course that was a long time ago, but even now I feel twinges of remorse remembering how those poor netted creatures all too often ended up succumbing to the ravages of life in a neglected jam jar near the back door. But happily this fish was different and he (or she, I suppose) lived with us for many years in a small heavily vegetated aquarium that my ever supportive mother had created especially for him.

My mum, with her enduring love of nature and of all its creations, was probably the other main inspiration who set me on track to become a professional biologist—but that's a story for another time. In any event, my tiddler spent most of his days lurking deep in the tank's vegetation so we called him Lurky. But at night he ruled the tank, swimming around with such grace and seeming enthusiasm that I would beg to be allowed to stay up and gaze into his mysterious world. Now, as a professional ichthyolo gist, I know that Lurky must have been a juvenile stone loach (*Barbatula barbatula* Linnaeus, 1758) and that, like most other loaches (close relatives of the carps and minnows), he was nocturnal, more or less, hunting and feeding at night on small crustaceans and insect larvae—or in Lurky's case, dried tubifex worms and fish flake. Perhaps he was my inspiration, I can't say for sure, but whatever it was, I have ended up spending my entire professional life exploring the world's waters and studying the remarkable fishes that inhabit them. All life depends on clean, fresh water, which plays such a crucial role in the maintenance of the health of our planet. And the fish I study can be seen as standing sentinel over those waters; rather like the miner's canary, they tell us a great deal about the health and functioning of the world's freshwater arteries.

From an ecological and evolutionary perspective it turns out that life in fresh water is very different from that on land, and there is something about these habitats that makes them so extraordinarily rich in life. Take those fish, for example; while freshwater habitats make up only about 0.8 percent of the Earth's surface, surprisingly there are almost as many fish species there as are to be found in all of the world's oceans—which occupy some 70 percent of our planet's surface. The reasons for such an exuberance of species in fresh water are many, but one of particular importance is isolation. We can think of fresh waters, whether they are rivers or lakes or sinkholes or swamps as islands—but inverse islands of water surrounded and isolated from one another by land. And just like terrestrial islands, which have spawned so many species in their isolation from mainland populations (think Madagascar, Galápagos, Borneo, New Guinea), freshwater systems seem also to be "species factories" where isolation has allowed genetic mutations to accumulate and local adaptations to arise, ultimately resulting in new species. And there certainly are a lot of them. Of the 53,633 living vertebrate species (animals with backbones—all the mammals, birds, crocodiles, lizards and snakes, tuatara, turtles, amphibians, and the fish) that have been recognized scientifically since the famous Swede Carolus Linnaeus formalized the process of describing species in the 1730s, more than half are fish. And as I indicated before, nearly half of those fish species are found in fresh water. It really is true to say that nearly a quarter of the world's vertebrates (that is, the freshwater fish) are found in less than 0.8 percent of the Earth's surface, and that's quite an extraordinary concentration of life—and a wonderful system to study and explore.

Exploration really is the best word. We know so remarkably little about so many of these aquatic species, and although I have cited exact numbers, the fact is that we still have no idea how many fish species actually exist on our planet, let alone where they all live or what roles they play in their ecosystems. Incredibly, between 200 and 300 fish species new to science are described every year. I stress new to science, because each one is usually well known to the people whose livelihood often depends on them. I work in the Congo River Basin—the freshwater heart of the continent. And fish are of tremendous importance to the millions of people who live alongside this extraordinary river, or who trade this compact protein source into the African heartland. In all the years I've been working in Africa, I have never been far from a river or from the fishermen whose lives are so intimately intertwined with it, and they have taught me so much more than any textbook ever could.

Fish provides more than 1.5 billion people with around 20 percent of their animal protein, and for many of the world's poor it contributes more than 40 percent. In Africa fish may often also form an underrecognized "buffer" that reduces hunting pressure on another alternative protein source—bush meat. Needless to say, the people who depend on fish know so much more about them than we visiting scientists do. Nearly always after I have walked into a small village seeking permission to collect fish in nearby waters, I have left accompanied by some of the most knowledgeable "ichthyologists" you could wish for. Always the local fishermen know exactly where the fish are, they know what they eat, where and when they breed, and they know how to catch them. They also usually have names for nearly all of them. But most of these fishermen are skilled at catching the big fish—the larger species that inhabit the main channels of the rivers—and these are usually sold or traded, and generally speaking are also known to the scientific world. In contrast, it turns out that most of the biodiversity in these tropical systems is actually sequestered in the small tributaries, streams, and backwaters that flow into the main river channels where the big fish rule. It is in these so-called marginal habitats that most of the new discoveries are being made, and here too we can rely on local expertise to help us find them. It is the women and children who usually harvest the smaller species, and recent research indicates that these small fish—African "tiddlers," I suppose—are the ones that usually end up in the family cooking pot and actually make up an important, if underrecognized, part of the rural domestic economy. So it turns out that in many parts of the world one man's dinner may often be another man's new species. On more than one occasion I have "collected" new species from the family cooking pot, and while working in Madagascar I even came across a new species slowly being smoked over a charcoal broiler in a busy marketplace!

Since I began my ichthyological journey, I have had the pleasure of describing more than 50 new fish species. Discovering and describing a new species is tremendously exciting and—to me, at least—great fun. It's certainly the closest I will ever come to immortality—because once you formally name a species that name will always be associated with the specimens described and the name of the describer. Of course, there are rules. In fact, there is a large legalistic compendium known as the Code (short for the International Code of Zoological Nomenclature) that formalizes the whole process, not coming up with the names themselves, but providing rules as to how the names can be constructed and applied.

At its most basic the Code specifies that the scientific name of a species is a combination of two names (a binomen), the first being the generic name and the second being the specific name. The generic name must begin with an uppercase letter, and the specific name must begin with a lowercase letter. Traditionally these two are always italicized. The Code further specifies that the original author and date of the description should be cited at least once on each work dealing with the taxon denoted by that name—thus the immortality. Every binomen has an authority, and that's why you will often see a species name followed by the name of its author. The common carp, *Cyprinus carpio*, was described in 1758 by Linnaeus so you will often see the name written *Cyprinus carpio* Linnaeus, 1758. The genus name, *Cyprinus*, is Greek, and although its derivation is unclear, Aristotle speaks of a fish with a soft, fleshy mouth, which he called κνπρινοζ —*cyprinus*. Some think the name may actually be derived from Cyprus (the suggested home of the goddess Venus—possibly an indirect nod to the high fecundity of the species), while the species name, *carpio*, is simply Latin for "carp." It's certainly not a rule that a species name must denote something biological about the bearer, but it is most helpful if it does. For example, a couple of species I described from a small lake in Cameroon had interesting "backstories," so I decided to refer to them with their names. Both belong to the genus *Tilapia* (an important group of food fish, sometimes called aquatic chickens, that are now a staple of aquaculture in many parts of the world). One of these species was strangely absent from the lake's coastal waters until the rainy season, when it suddenly turned up in large numbers. The local people referred to it as the rainy season fish, because that's when they caught it. So I came up with *Tilapia imbriferna,* from the Latin *imbrifer* (rainy) and *ernus* (pertaining to time). The second species was known only from the deepest parts of the lake, so that one became *Tilapia bythobates,* from the Greek *bythos* (the deep) and *bates* (walker or goer). The tilapia of the rainy times and the deep-walking tilapia—I am still rather proud of those two.

Discovering and describing species is one thing, but making sure they and their habitats persist into the future is quite another. In our world of burgeoning human population, climate change, and water scarcity, the increasing threats to fish and their fragile aquatic habitats have led me into the realm of conservation biology. Fresh water is of such tremendous importance to our species, as it is for all life on the planet, that it has been said that humans are simply an invention of water to get itself moved around the planet. And a quick look at some statistics certainly lends credence to that position. Today more than 60 percent of the world's total stream flow is regulated, and dams and reservoirs now hold back on land about 4,000 cubic miles of water—that's five times the volume of all the world's rivers into the oceans. There is now so much water impounded by dams in the Northern Hemisphere that it has actually altered the distribution of weight on the planet, resulting in small but measurable changes in the Earth's rotational and gravitational fields! That's how large our impact has been, and in many places we have completely transformed the aquatic environment.

In the face of such tremendous changes, it is not hard to imagine the impact that our species' use of water is having. I have described more than 50 new fish species, but to be honest, I am not exactly sure how many of them are still around. Threats are obviously greatest when a species has a limited geographical distribution—if something happens to its habitat (a small river drainage or isolated lake), the potential for extinction is much higher than if it occupied a wide geographical range. Many of the new species that I have described occupy such habitats, places where a small water diversion to irrigate crops, or overfishing in a small lake, could mean the loss of an entire population or even species.

When I started working on my Ph.D. thesis, I was driven by academic interests and by trying to understand how diversity is generated in freshwater systems—interests that have often taken me a long way from the riverside and into the high-tech world of the genetics laboratory and to many professional meetings, college campuses, and seminar tours. But really, ever since my first visits to Africa (where the fish that started me on this journey live), it has been clear that however interesting and important an understanding of the generation of diversity is, how it is to be maintained is in many ways even more important.

So since those early days I have always tried to link up with conservation organizations and local institutions in the countries I have worked in, to try to see how my findings and data can be used to help inform the local people and their government representatives about the importance of conservation and sustainable use of their natural resources. Of course, working with fish gives me the obvious advantage of a clear link to human livelihoods through their role in the food chain, and as indicators of water quality and health, fish are often of interest even to landscape managers and bureaucrats. But beyond these obvious links I have found that there really is something to be said for the old adage *scientia potentia est* (knowledge itself is power). To know that your rivers and your fish are something unique in the world and a part of your nation's patrimony sometimes really can have resonance. This was brought home to me on a recent trip to my study site on the shores of the Congo River, near the small settlement of Luozi about 60 miles downstream of the megacity of Kinshasa, the capital of the Democratic Republic of the Congo. Getting to Luozi each year is an adventure in and of itself. Making it through the security checks, overcrowded flights, international connections, and frequent delays is tough enough, but doing it with the inevitable overload of excess bags full of camping supplies and heavy scientific equipment is positively nerve-racking. Arriving at Kinshasa's chaotic N'Djili Airport, sleep deprived and with nerves jangling, is a mixed pleasure, but with patience, lots of smiles, and the invaluable help of our field partners from the University of Kinshasa and the local World Wildlife Fund office, we always seem to make it through to emerge dazed and punch-drunk into the rowdy Kinshasa night.

After a couple of hectic days spent getting the trip's logistics on track (the best laid plans . . .), buying supplies, meeting up with old friends, and checking in with our collaborators, we are ready to head out to Luozi. But getting around in the Democratic Republic of the Congo is not easy. Years of conflict and neglect have left the country with very little in the way of working infrastructure, and its road network is among the worst in Africa. In fact, there are so few paved roads in this massive country that it has been estimated that there are only 20 miles of paved road per million people. I mentioned that our field site at Luozi is about 60 miles from Kinshasa, but getting there takes a full day of arduous 4x4 driving, much of it on a rough, dusty, potholed "road," and with the last leg on an ancient rusty, very nonconforming ferryboat that disgorges us at the foot of the small fishing settlement of Luozi. There I have been working with members of a local fishing cooperative, and together over the past three years we have tried to document and research the extraordinary diversity of fish species of this region. By now many of the fishermen are well known to me, and my arrival is marked by a lovely ceremony of gift giving. On this last trip my gifts to the cooperative were a small digital camera and a six-man tent—and I got a goat and a huge sack of dried beans. The goat stayed behind to live on another year, and back in New York City I am still working my way through those delicious beans.

Our work in the region has more than academic implications, as this part of the Congo River has the greatest hydropower potential on the continent. From Kinshasa down to the Atlantic Ocean, a short stretch of about 250 miles, the river drops some 920 feet in elevation, resulting in some of the largest and most extreme rapids on Earth. It is the river's hydrological complexity that is probably driving the isolation of fish populations, and this in turn is very likely the reason there are so many fish species found here. But the system is also a potential gold mine in terms of hydropower, and there are plans to expand and develop a series of large dams downstream of Luozi that will harness the river's power. The need for reliable and sustainable energy is nowhere more pressing than in a country like the Democratic Republic of the Congo, where most people still rely on the environmentally destructive and unsustainable use of charcoal for most domestic needs. It is hard to argue against development. But along with the millions of people who depend on the fish of this free-flowing river, I too am concerned at what the many potential downsides of damming the Congo will be. The looming uncertainties of these changes hang heavily on the minds of my friends and colleagues who live alongside the Congo. But all agree that to begin to fully understand the potential ramifications of damming the river, we need, at the very least, to know what fish live there and how they use the river's many habitats—so here in the Democratic Republic of the Congo, in a very real sense, *scientia potentia est.*

When this year's field season ended and I prepared to leave, Thomas, the head of the fishing cooperative, came to me with a request. Not a request for money, or for new fishing gear, but a request for a book—a book that would illustrate and give the scientific name for each of the hundreds of fish species that make this stretch of river unique.

These fishermen know the river and its fish far better than I ever will, yet through our work together they have come to understand that their river is special and something to be proud of in a new way. And that makes me feel really good and a little more hopeful for its future.

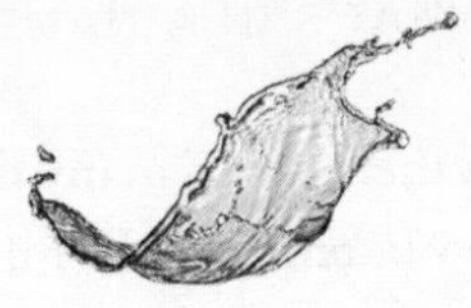

THE FEAR OF DOING WITHOUT

Ellsworth Havens

Ellsworth Havens, a native of Brooklyn, New York, earned a master's degree in public health from the University of Tennessee and a master's in business administration from Fairleigh Dickinson University. He has devoted much of his career to helping plan and regulate pediatric health care centers, rising to senior vice president of Englewood Hospital in Englewood, New Jersey, and chairman and board member of the New Jersey health care regulatory and planning board. He has also served as president of the American Marketing Association, New Jersey chapter, and chairman of the Educational Opportunity Fund board for the New Jersey Commission on Higher Education. Since joining Rotary in 2006, Havens has been involved in water issues, serving as the chair of the water, health, and hunger committee for his local club as well as chair of development for the water and sanitation action group for Rotary International. He is president of the Madison, New Jersey, Rotary Club.

As a boy growing up in Brooklyn, New York, I knew about having to do without. There were many times in my youth that I went to bed hungry. In my neighborhood most families struggled, although living standards varied from apartment building to apartment building. I knew that compared with others outside my neighborhood, we were poor. The families I considered rich were, of course, not rich in hindsight, but compared with what I had, they seemed to be at the time. Of course, food wasn't the only thing that my family and I did without; there were few clothes or toys, no money for school trips, and always the fear of being evicted. But I always saw those things as temporary pains and fears.

What was always my worst fear was not having any food. It was a fear for myself and for my family. I knew that the quantity and quality of clothes and toys naturally varied from family to family, but food was different. Many nights I stayed up, afraid to go to sleep, worrying about how and where I'd get my next meal. What got me through many nights was thinking about the faraway places and people I'd read about. This helped me forget being hungry. And I knew that in some of these faraway places there were people even worse off than I was.

My personal journey has taken me out of those conditions, but knowing and remembering that fear has never left me.

My first graduate degree was in public health, from the University of Tennessee. Most of the course work dealt predominantly with public health issues in the United States, but the core courses dealt with classic public health issues of disease and prevention, more relevant to conditions in developing countries than anything I'd experienced. Little did I know at the time that someday I would be confronted with those conditions.

Three years ago a friend of mine asked me if I had ever thought of becoming a Rotarian. I had spent a good deal of my professional life serving on various nonprofit boards and state boards and commissions but had never given any thought to Rotary. In fact, I knew nothing about Rotary.

After attending a couple of meetings I decided that it was an organization that I would like to be a part of. Since 1988 Rotary had focused on eradicating polio. With that goal almost achieved, the global crisis of potable water and adequate sanitation was moving to the forefront.

I volunteered to work on Rotary's water and sanitation project. Suddenly, for the first time in decades, I was reading about diseases and conditions I had studied in graduate school, including the devastating health and social aspects of unsafe drinking water.

I had spent a good deal of my career either planning, regulating, or developing pediatric health care centers. I opened advanced neonatal intensive care nurseries and started specialty surgical programs. I was always amazed with advances in technology, technique, and medicine that had evolved over time.

As I learned more about water and sanitation internationally, I was confronted with the reality many people face: the choice between drinking dirty water or dying of dehydration. Most choose to drink, which is why water-related parasites and diseases are the leading cause of death in children—a sobering fact. I soon began communicating with Rotarians in India, Central America, and Africa, asking questions and seeking advice.

Attending my first Rotary annual meeting helped me get a better understanding and clearer picture of the crisis and all of the efforts being made to solve it. Like any crisis solving, it takes the effort of countless people in many ways. Rotary had recognized that an expanded organizational emphasis was going to be necessary.

In 2007 the Rotary International Board had approved establishing a nonprofit action group dedicated to the crisis of water and sanitation, which works in tandem with the existing network of Rotarians already involved in raising money and implementing projects throughout the world. For me it offered a way to address the fear I knew as a child by alleviating the suffering of children with similar fears. Clean water, like food, is something no child should live in fear of not having. And no child should have to drink water infested with deadly bacteria.

One country in which too many children die of diarrhea and waterborne communicable diseases is the Dominican Republic. There are nine million Dominicans and one million Haitians living in the Dominican Republic, and it is estimated that more than 25 percent do not have regular access to potable drinking water.

During a trip that I took to the Dominican Republic in 2007, I witnessed the day-to-day reality of the situation. Before I'd left, I had met with Rotarians who discussed with me the plight of the country, but it wasn't until I was actually there, and when I talked to our tour guide—a local man—that it became personal. A father of five, he lived in Santo Domingo. I asked him about his family and how they use water. He explained that they have access to water for cleaning but not for drinking; they must buy bottled water. I asked him what happened when he couldn't afford to buy water. His response was that they pray a lot.

Following this conversation, it was difficult for me to focus on sightseeing, because at every stop, the vendors selling water to the tourists reminded me that behind the walls and down the streets there were locals without any water. And even when I went back to our resort, there was a new supply of bottled water for us, served by workers who most likely didn't have clean water in their own homes.

Of course, it wasn't just a problem in the city. All the workers with whom I spoke were friendly, and many of them had friends or relatives who lived in New York or New Jersey, so it was easy talking to them about Yankee or Boston Red Sox baseball. We also talked about water: They said that though they served it at work, they never expected to have it at home. Some lived in smaller towns or rural villages. They just accepted the fact that if you were a tourist or a rich Dominican you had clean water, and if you weren't, you didn't. I continued to think about their comments every morning when I saw the water truck bringing clean fresh water for the tourists at the resort.

I met with members of a Dominican Rotary Club, and together we went through many scenic areas. Unlike the city, which had plumbing, most of the countryside did not, so there wasn't even direct access to dirty water.

Returning to the States I was more convinced that perhaps the efforts of even a few of us actually could benefit countless people, over time. Together, we could move mountains instead of molehills.

As I got more involved, I spent more time talking with various groups, including other Rotary clubs. I realized that more information and education is clearly needed. I'm continually questioned about cost and technology. My organization has learned through experience that local needs and involvement should drive the selection of appropriate technology, based on lots of factors including cost. For example, if properly installed, a rain-harvesting system of gutters and collection costs relatively little, as do hand-drilling wells and boreholes, simple technology like the LifeStraw mobile water purification tool, hand-powered pumps, ceramic and biosand filters, solar purification, and chemical purification and separation. These all work out to pennies per year per person, which may still be cost-prohibitive for recipients, but are inexpensive for donors. The challenge is not the cost of the technology but involving the local community in selecting the correct approach. What's important to keep in mind when implementing a project is that expensive or inexpensive, a technology that is not sustained because of local customs, mistrust, disinterest, or other factors is ultimately useless. It is critical for the local community not only to be involved but to assume ownership of the project

The good news is that through Rotary and other organizations the problem is being addressed. The biosand filter, which costs about $60, was invented in 1990 by an engineering professor from Canada. It is believed to reduce waterborne pathogens by more than 90 percent. The filter has no moving parts, consisting only of a cement or plastic container filled with sand and gravel.

Water is poured in, and as it filters through the mechanism, beneficial bacteria growing on the sand kill microorganisms.

For seven years now, the Rotarian-led Children's Safe Water Alliance has installed filters in the Dominican Republic. Nineteen thousand filters have now been installed, providing clean water to an estimated 100,000 residents. That effort has been supported by more than 200 Rotary clubs in the United States, Canada, and the Dominican Republic and other Caribbean countries.

But the work is far from done. That still leaves millions in rural and urban areas without clean water. The dilemma of having to buy bottled water, which is not always clean, or not having any available supply, still exists throughout the country.

In addition to our work in the Dominican Republic, we've also worked in Ghana and the Philippines, bringing fresh water to impoverished areas. We hope that these pilot projects will be just the start.

In fact, there is pending legislation in the U.S. Congress that would expand the U.S. Agency for International Development's scope in water and sanitation projects. In addition to these efforts, we continue to raise and donate funds to projects throughout the world. Working with other NGOs, corporate partners, and alliances, and through our members every day, we are making sure a child or a family will not have to live in fear of not having clean water. But we've only just begun.

I find myself almost every day talking to a group or an individual about the crisis. Fortunately, the vast majority of Americans have never really suffered or lived in fear. Even in depressed urban areas and rural America, inadequate sanitation is unknown. If there is a water crisis in the United States, it is not a health crisis but one of limited use. In my area most of the debate relates to how often you can water a lawn or wash your car, so getting people to truly understand the issue is challenging. But what I tell them is to picture one of their children or relatives lying in a bed, dying, because they don't have a glass of clean water. For most, it helps them understand.

As for me, I haven't had to fear hunger in decades. Working with a million other Rotarians to save a child and family from similar fears helps me sleep a little better.

WATER MAN

Rajendra Singh

Rajendra Singh is known as the Water Man for his role in restoring water to arid desert villages in Rajasthan, India. He founded the water conservation organization Tarun Bharat Sangh in 1984, through which he has mobilized rural communities to revive traditional water-harvesting structures in more than 1,000 villages, thereby regenerating 2,500 square miles of land. Over a period of 15 years, these villages stayed drought free while the rest of Rajasthan and India experienced droughts caused by decreased monsoons. Singh also played a pivotal role in revitalizing five shrinking rivers in Rajasthan; they now flow year-round. In 2001, he won the prestigious Ramon Magsaysay Award for community leadership for his pioneering work in water management.

I WAS BORN TO A FAMILY OF FARMERS in India's Uttar Pradesh state, a populous region bordering Nepal. My village was surrounded by canals that diverted the waters of the two mighty rivers of India, the Ganges and Yamuna. In this fertile land, we never lacked water for irrigation; our fields were green and lush. I would soon discover that other places in India were very different.

Every morning before going to school, I used to repair the boundaries of our farms and plow the fields. But my father never wanted me to be a farmer. Instead, I studied the Ayurveda system of medicine and joined the Ministry of Youth and Sport as a national volunteer, which took me to the city of Jaipur in the state of Rajasthan. Covering more than 10 percent of India's area, Rajasthan is just to the west of Uttar Pradesh, but it lacks that state's wealth of water. In fact, Rajasthan is well known for its recurring droughts, its arid and semiarid land possessing only about one percent of India's total surface water.

After a year in this destitute region, I quit my job and decided instead to work to bring dignity and prosperity to this part of rural India. Discouragement came quickly: My plan to start a school for the children of Rajasthan's Gopalpura village met with no response from the villagers. Depressed as I was, I wasn't willing to give up. Seeing my desperation, Maangu Meena, a wise old man of the village, consoled me. "You have not understood what is needed here," he told me. "We want *water* first. You need to build a *johad* [a traditional rainwater-harvesting and storage structure] so that the water does not run away, but is held back to percolate into the ground." With several other volunteers, I founded a voluntary water conservation organization, Tarun Bharat Sangh (TBS).

Soon the first small johad was ready. In the very next monsoon the villagers saw how rain falling from the black clouds ran into the johad. From there it percolated into the ground, adding to the water in their wells, fields, and pitchers. Birds sang in the green hills as the gift of water changed their ecology and their economy.

For me, the simple act of building johads became a way to develop poor rural regions—a way to revive an area's ecology for better food production, vegetation growth, and river restoration. In the past 25 years, we have seen more than 10,000 rainwater-harvesting structures restored, and the effects are visible. Wells and aquifers are being recharged. Rivulets that had been dry for many years are flowing again. Agriculture production is increasing.

Besides encouraging water security through johads, my organization has encouraged villagers to plant crops, like grains, pulses (such as beans), or animal fodder, that use less water. Village people started using compost manures and local original seed for their farms, reducing their dependency on expensive fertilizers and commercial seeds.

Water conservation has even helped to reverse the emigration of villagers. With more animal fodder available, they have been able to sell milk products through an informal cooperative arrangement. But women have been the primary beneficiaries of the regeneration of water sources and forests. In this region, women bear the brunt of physical labor, carrying fodder, fuel wood, and water—"women's chores" that consume, on average, 18 hours a day. With the creation of rainwater-harvesting structures, some of their backbreaking work has eased and they now find time for themselves. Girls who once had to devote so much of their time to fetching the family's water are now able to go to school.

This movement was hardly a cakewalk. After I began, government officials started sending me notices saying that the johads were illegal because they were built without government permission. To every notice I replied that water conservation is not a personal matter—it benefits the community, agriculture, cattle, wildlife, and Mother Earth. It is work that the government should be doing but has not. Therefore, I took this responsibility upon my own shoulders, in partnership with the local community.

However, even as communities were restoring their johads with the help of TBS, we could not help seeing that the hills around us were being gouged mercilessly by mining, affecting their water and drainage. We realized that even the hills needed protection. Our campaign against mining in the forest areas eventually resulted in the mines' closure. During this struggle, TBS activists repeatedly faced the wrath of the mine owners; even I was attacked three times.

I'm often called the Water Man or the River Man, but only recently have I come to realize that water conservation is important even in my own, well-watered native region. Three years back, I was traveling home with my son. While crossing a bridge over the River Yamuna, I was taken aback by its condition. I had spent my childhood playing on its banks, but today, even in the country's capital of New Delhi, the river is like a drain carrying black water full of industrial effluents and sewage. My son asked me, "Will I be able to see this river as it was in your day?" When I didn't answer, he asked, "Why do people call you River Man, when your own native river is dying?" I told him that people call me River Man because my water conservation work resulted in the rejuvenation of five small rivulets in dry regions. This didn't satisfy him, and he shot another question at me. "Why do you work only with rural communities, and not with urban people who are killing these rivers?"

This question was a challenge. It was true that even our most revered rivers were being abused and forgotten. People think of them only as ways to transport garbage, effluents, and sewage. On the day after my conversation with my son, a young activist came to me and asked, "What are you doing for the Yamuna River?" Commercial interests, he told me, were eyeing the land around the river; if we didn't stop this development, nothing would be left, and the next generation would see it only as a line on old maps. So some of us decided to start fighting back by planting trees along the Yamuna riverbed at the site of the commercial encroachment. Other Delhiites joined our effort, and soon we had planted about 150 trees, including *kadamb, gular, peepul,* banyan, and *amla* trees. These species, which once grew on the banks of the Yamuna, prevent soil erosion, control silting, help recharge groundwater, and contribute to the perennial flow of the river.

Soon the police came and stopped us, but that didn't slow the launch of the Save Yamuna campaign, dedicated to preventing the river's exploitation and pollution. After two years of effort, the government listened to us and halted construction in the riverbed. This saved around 25,000 acres of land around the river from encroachment.

Now the biggest challenge in front of us is to restore the clean flow of our great rivers. For instance, we are trying to set up talks with central and state governments to set a policy for the Ganges (Ganga), hoping to make it free of pollution, exploitation, encroachment. One of the greatest rivers of the Indian subcontinent, and the source of life for hundreds of thousands of people, the Ganges is also one of the rivers most threatened by climate change. It is fed by the Himalayan glaciers, but rising temperatures mean that many of these glaciers are melting fast. The water they supply could diminish significantly over the coming decades, with catastrophic results. In the long run, the water flow in the Ganges could drop by two-thirds, affecting more than 400 million people who depend on it for drinking water.

Other activists and I are organizing street plays, conferences, seminars, and awareness programs in schools and universities. Learning from the Save Yamuna campaign, we started the Save Ganga campaign. As a result, the Ganges has recently been recognized as a National River by the Indian government. The government also established the National Ganga River Basin Authority to coordinate planning, financing, and monitoring of the river. On the 62nd anniversary of India's Independence Day (August 15), we celebrated Independent Rivers Day. Volunteers from different parts of the country participated and took an oath to devote the rest of their life to protecting and saving rivers.

Freshwater resources and ecosystems are under great threat from climate change and many non-climate-related demands. Water is our life; it should not be seen as a commercial commodity. Our natural resources are part of our culture, civilization, religion, and traditions; they should be above cost-benefit analysis.

We should also recognize that small things are beautiful. Small rainwater harvesting structures, controlled by the community, can revive the economy of poor villages. On the other hand, big dams, once considered temples of development, are destroying and displacing communities.

Water conservation is not about scarcity, but rather about water's careful use and equitable, distributed access. We need to rework the ways in which we manage water, so that local water resources foster inclusive growth at all levels. Rivers have nurtured civilizations since the beginning of time, and people have always respected and revered them for providing the essence of life. Whether it is the Ganges or the Nile or any other river in the world, a river is like a nation's veins. Its health symbolizes the health of the people.

A RIDE FOR NATURE

William "Waterway" Marks

William "Waterway" Marks was born in Morris County, New Jersey, and raised on an organic farm. He studied at the University of Corpus Christi and at Fairleigh Dickinson University, from which he received a bachelor of science degree. While at Fairleigh Dickinson, he investigated fish kills and industrial pollution; his water-testing and photographic evidence led to the federal grand jury indictment of a polluting industry. Marks is the founder of Martha's Vineyard's first state-certified water-testing laboratory, and the founder of Vineyard Environmental Research Institute, which has received acid rain research grants from the U.S. Environmental Protection Agency and the American Water Works Association. He was elected to the board of the Massachusetts Association of Conservation Commissions, which helps protect wetlands in more than 300 communities statewide. He is an award-winning author of *The History of Wind Power on Martha's Vineyard* and *The Holy Order of Water: Healing Earth's Waters and Ourselves.* He published and edited *Water Voices From Around the World.*

THE HEAVILY LADEN BUCKSKIN follows closely behind the lead horse as we climb the crest of a mountain in Cleveland National Forest. Looking back over my right shoulder, I squint against the bright light. It is a clear, crisp January evening illuminated by a shimmering sun reflecting off the silver surface of the distant Pacific Ocean.

The horseback journey has begun. It is a journey that will take almost two years and traverse more than 7,000 miles. A journey I have entitled Ride for Nature. A journey during which I anticipate living outside day and night as I cross America in the fashion of our Founding Fathers and Mothers.

We are embarking on an uncharted zigzagging trail from the coast of San Diego to the coast of Maine. The trail is destined to involve us in many adventures as we cross paths with hundreds of people.

We are to experience glorious days and some that are horrible; suffer the hot and the dry, as well as the chill of rain, snow, and hail; gasp our way through a sandstorm and push our way over dunes riddled with potentially crippling holes from rodents; accidentally cross Yuma Proving Grounds during the testing of Star Wars laser weapons; experience Native American reservations and alternative-lifestyle communes; ford rivers; work our way across oil fields and farms; earn a few dollars by rounding up cattle, training horses, shoeing horses, and appearing before audiences; navigate through cities during rush hour; and cross the George Washington Bridge and ride through the South Bronx during an electrical blackout accompanied by looting and fires.

My heart, mind, and body are ready to embrace the adventure of exploring America like our ancestors. As well, I am aware that the trail will bring a palpable consciousness of water: water for the health and survival of the horses and myself; water as it interacts with species of life that define our landscape; water that has faithfully served as the economic foundation of America.

Just three months ago, I was sitting, dressed in a three-piece suit, in an air-conditioned office on the fourth floor of Newark's City Hall. My job as senior environmental analyst was to manage most of the city's water issues. My labors included responsibility for the city's drinking water, polluted runoff from Newark's International Airport, storm water runoff from streets, and the waste dumped by industries and ships.

While employed by Newark, I also enjoyed a one-third scholarship with the University of Michigan's School of Natural Resources. As the first ecologist in America to be hired by a city, I was being granted life credits toward a master's degree by the university.

I was overflowing with enthusiasm. Surely, I thought, I can help bring about a positive change in this dying city. This enthusiasm was fueled by my success in college at finding the cause of fish kills and investigating industrial polluters along rivers.

All was going well with my job until one day, while at a meeting with the Port Authority of New York & New Jersey in the World Trade Center's Gold Room, I was visited by a vision: A man was riding a horse down the side of a mountain toward a river. This vision kept returning for several months—usually in the middle of a high-powered meeting.

After this vision, upon awakening one morning, I decided to quit my job. When I told my boss I was resigning to ride a horse across the United States, he thought I was joking and laughed. When he realized I was serious, he gave me 30 days' paid leave to reconsider. Before those 30 days passed, I had sold all my worldly possessions and was on a plane to San Diego. When my brother Lefty met me at the airport, he asked, "Where is your luggage?" I reached into my pocket and produced my two pieces of luggage—a toothbrush and a checkbook.

As the arching sun kindles blazing heat, I seek protection for the horses. The Southwest is suffering one of its worst droughts in recorded history.

My original plan was to ride across southern Arizona and New Mexico and then drop down into Big Bend National Park in Texas. This plan was abandoned after I visited a rancher who had lost hundreds of cattle and sheep.

As I was approaching a ranch, vultures were circling above. Soon I saw a macabre scene. A large group of vultures were squawking, fighting, and pecking away at the bloated carcasses of dozens of cattle and sheep. The smell of dead animals spooked Shalamar and Buck, forcing me to ride with tight rein and taught lead rope.

Some cattle carcasses had bloated bellies. Others were picked clean, with just the ribs arching skyward. The dead sheep appeared as protruding puffs of dirty brown wool sticking out of drifting sands.

The two dogs protecting the weathered homestead warned of our presence from a mile away. The rancher, with a gun strapped to his hip, waited for me to dismount. "Howdy. I'm Steve Bains. How can I help ya?" Under the shadow of his stained white Stetson, Steve's deeply wrinkled face looked like crinkled leather; his thickly callused handshake felt like a vise.

It turned out that, at the age of 76, Bains was preparing to leave his ranch of more than 30 years for good. The loss of his herd to drought had done him in. His heart and spirit broke as he watched his cattle and sheep wither away and slowly die.

It was ironic to see water filling Steve's bloodshot eyes as he shared his sad story.

"In the beginnin', the range grass dried up and went to almost nothin', forcin' the herd to travel long distances to find somethin' to eat. Since I got only one well, the herd was forced to travel longer and longer, back and forth between food and the water. Some of the sheep and cattle started to drop dead. I tried feedin' 'em straw with liquid vitamins and minerals sprayed on, but it didn't help. I watched many of 'em lie down and take their last breath. Tried nursin' a diein' calf; it died in my arms. Hate to tell ya—but me an' the dogs been gettin' by by eatin' cooked meat from some of the fresh carcasses."

The drought had forced change into Steve Bains's life. He was being forced to migrate to a place with water.

Over millennia, the cyclical migration of humans away from drought-stricken regions was normal. However, with today's political boundaries, such migrations are difficult, if not impossible—thus creating millions of environmental refugees in regions around the world each year.

I shed a few tears as Steve's lean body, covered with dirty clothes, stood and watched my horses drink water from his ranch's watering trough. In a way, I could tell it made Steve feel proud that he was able to help a stranger and his horses survive.

As we made tracks across the dusty, lifeless land away from the ranch, the horses and I headed north—north, where it was cooler and wetter and where the grass was greener. I wanted to avoid the risk of the horses or me becoming food for vultures.

As blackness thins to dull gray, I spy the first faint glimmer of light to the east. Slowly, it looms larger and more intense—giving us fair warning of the coming heat. Traveling since early morning under the refracted light of moon that illuminates our trail, we are standing in the middle of the Sonoran Desert. The nearest house or road is at least a five-days' ride.

To protect my eyes, I pull the dusty brim of my cowboy hat down low on my forehead. In the distance, the cool, dark shadow of a mountain slowly yields to the hot embrace of the rising sun.

Carefully, my horses and I pick our way through the intertwined green arms of prickly mesquite and cactus as thorns threaten anything that moves in this land of deadly sand. With each movement we do our best to survive this potentially dangerous day—a day that promises no rain, a day that threatens to take our lives. With every movement and exhalation, we lose valuable water as arid air sucks moisture from our bodies and lungs.

My lead horse, Shalamar, and packhorse, Buck, lick the air with swollen tongues—it has been two days since they tasted water. If we don't find water today, there will be hell to pay, I say to myself.

Now, after many days and hundreds of miles, my burned face is sore and lips cracked. My once manicured hands are dry and swollen with new calluses that bleed at the finger joints. To escape the midday heat from above, and the 150-degree sands below, the horses and I need shelter.

A short distance ahead stands a cluster of magnificent saguaro cactuses with their thick trunks and arms silhouetted against the rising sun. Before long, I have Shalamar and Buck standing in the cool shade of two tall saguaros. Selecting a nearby mesquite tree, I toss a piece of canvas over its low branches. A resting rattler gives warning. I say, "Oh, I'm sorry," and then slowly remove the canvas. I throw the canvas over another nearby mesquite, check the ground for rattlers and scorpions, and place my wool blanket on the ground.

Tucked under the protective cover of the canvas, I am at least 20 degrees cooler. I look over at the curled-up rattlesnake. It lies motionless, as though near death. Like the rattler, the horses and I rest quietly and wait for the heat of day to fade away.

Along the trail we've seen and met many rattlesnakes; they are usually respectful, and when given space, just slither away to live their day. On some mornings after a cold desert night, I've seen close to ten rattlesnakes stretched out straight as they blissfully absorb the sun's warmth.

After a long snooze, I write in my diary with a hand so soiled that it leaves a dirt trail across the page. As I write, lizards scurry over my sitting blanket. They, too, enjoy the cool of my impromptu shade.

I am hungry but choose not to eat the granola in the saddlebags. Digesting such food would rob my body of precious water.

Due to circumstances, the horses and I are forced to endure a food and water fast. The three of us have been reduced to nothing more than muscle and bone.

Just thinking of water intensifies my thirst. I wonder if Shalamar and Buck also think of water. Surely they must.

As the hours pass, I meditate on my surroundings as the shadows of the saguaros shift like those of a giant sundial. Being careful not to be caught sleeping out of time, Shalamar and Buck move their standing positions to stay in the shadows.

It is hard to imagine that some of the saguaros I am gazing at may have existed as youngsters during the days of Cochise, the great Apache chief, days when the Apache wandered freely over tens of thousands of square miles as though it was their familiar backyard.

Looking at a massive saguaro often commands one's attention. Saguaros stand as the universal symbol of the American West and are oftentimes featured in movies depicting desert scenes. As well, because of their human appearance, they are often referred to as silent desert sentinels. Besides standing as the largest life-form in the desert, these subtropical giants can live for 200 years and reach heights up to 75 feet.

The saguaro's accordion-shaped body allows the trunk to expand and contract with available water. When times are good, the saguaro's body will expand and contain up to 90 percent water. When times are bad, its body will contract and lose up to 60 percent of its water yet still survive.

When wild horse trails guided us into Wickenburg, Arizona, the Dude Ranch Capital of the World, we passed the largest saguaro I've ever seen near the Hassayampa River. We literally rode south down the middle of the river and its meandering chilly snowmelt waters for more than 20 miles. The riverbed took us through high-walled canyons gracefully carved over millions of years. Eventually we came to a spot where the flowing river water vanished into the desert sand. We followed the dry riverbed to its juncture with the Gila River. Along the ancient Gila Trail we passed several small groves of saguaro.

Like many river travel ways around the globe, the Gila River continues to serve as a transportation route that dates back to ancient times. Archaeologists estimate that human interaction and nomadic occupation along the Gila dates back more than 20,000 years. Skeletal evidence and trading items such as human-worked shells, stones, and metals give testimony to active prehistoric trade, hunting, and travel along the Gila River Valley.

The Gila Trail enjoys a storied existence due to its westerly flow from the highlands of New Mexico to the Colorado River on California's border. When a Dr. John S. Griffin traveled the Gila Trail in the mid-1800s, he described the territory with these words: "Every bush in this country is full of thorns . . . and every rock you turn over has a tarantula or centipede under it . . . the most beautiful specimens of rattlesnakes . . . lizards and scorpions . . . hard to imagine anything so barren."

While traveling the backcountry along the Gila River and other Southwest waterways, I discovered prehistoric rock carvings known as petroglyphs, and paintings called pictographs. Such prehistoric art may be found near special sources of water on every continent.

The hundreds of primitive art forms I saw depicted water symbols and maps of waterways; stories of weather patterns; fluctuations of river levels and rainfall; calendars for planting tied to solstices and equinoxes; wild and domesticated plants; families, warriors, and ghostlike figures; horses, coyotes, birds, and snakes; weapons; hunting adventures; wildlife tracks; geographic features; musical instruments; cosmic entities such as Venus, Mars, the sun, moon, and comets; imprints of human hands—and many other things.

Similar examples of the artistic presence of humans have been documented along the Tigris and Euphrates Rivers, the Nile, the Amazon, and many others. The story of humankind is truly writ in water.

The proximity of creosote bushes, mesquite, and paloverde trees to the two saguaros shading my horses is no coincidence. These plants serve as "nurse plants" to the fragile young saguaro seedlings, which require protection in their early years. It takes several decades before their skin is thick and tough enough to deal with desert sandstorms, cold, direct sunlight, and searing winds.

Besides surviving these threats, saguaros also experience threats from well-intentioned human management. When traveling north of Yuma, Arizona, I met a national park ranger who shared his story about a rattlesnake experiment.

Apparently, someone in the upper echelons of management was concerned about the number of rattlesnakes inhabiting Saguaro National Park. The basic fear was that visiting tourists hiking sections of the park's 165 miles of trails had little awareness of rattlesnakes. After several meetings, the executive decision was made to eradicate large numbers of the snakes.

Within a few years of eradicating the rattlesnakes, park ecologists began noticing the rapid disappearance of young saguaro cactuses and other plants throughout the region. Further observations revealed how the removal of rattlesnakes caused a population explosion of rodents such as rats, mice, and rabbits. Without the predatory control of rattlesnakes, rodents were free to reproduce and feast on the fragile young cactuses as a source of water and food. Needless to say, rattlesnakes were quickly introduced back into the ecosystem of Saguaro National Park.

This story is similar to others we see in ecosystems around the globe. Story after story tells of how humankind's disruption of water flowing through a species in a food chain can quickly have impacts on an entire ecosystem. As the original source of life, water fulfills a vital role in the food chain. If we disrupt or alter water's energy in any ecosystem—by changing the course of a river, building a dam, polluting, or eradicating a species—we weaken our planet's living fabric.

As the sun's energy ebbs and the desert cools, I watch the nearby resting rattler quietly slither away into the undergrowth. Rattlers will instinctively avoid being around humans, horses, cattle, buffalo, and other animals whose earth-thumping feet send warnings.

It is time for us to make tracks. Before saddling up, I look out across the expanse of desert and whisper a prayer for finding water. From a nearby mesquite tree, a barking round-tailed squirrel stands tall on a branch as it seemingly chatters goodbye.

We head south toward a gentle, sloping valley. The desert dusk thickens as the sun kisses high thin clouds with a blending palette of red, orange, yellow, and pink. A series of sandy gullies lead us into the valley as a cool low-lying mist materializes and surrounds us.

Off in the distance ahead we are greeted by the long, spooky howl of a coyote. This howl is answered several times from different directions. When the howling ceases, there is only the sound of our footsteps and breathing. With each step the arid sand grains emit a squeaking tone from friction caused by my boots. Whenever we stand still, the silent world surrounding us also stands still, as though patiently waiting for us to make a fatal mistake.

Again, the silence is broken by coyote calls. This time, the howls are bunched together to our right. We stand motionless. A rising, thick fog makes it impossible to see the coyote pack. The howls grow louder and approach closer and closer before veering off behind us and fading away into nothingness.

As we continue, the mist thins and disappears.

Travel during twilight hours is risky, but it helps us to retain our body water.

My kidneys will filter up to 40,000 glasses of water a day. When dehydrated, my body hoards water as my kidneys and liver work overtime to clean and recycle whatever blood water is left.

My light brown cowboy hat, khaki-colored wool shirt, oversize silk neckerchief, and goose-down vest serve as my desert attire. The wide-brimmed hat has vents to help my head and brain stay cool. I know that staying calm and keeping my head under cover helps me to think clearly.

As we make our way through the expanse of desert with its after-sunset ambient radiance, I keep a keen eye out for animal tracks and patches of green that may help lead us to water. After two hours there is nothing—not even a hint.

To minimize stress on the horses, I walk instead of ride. This allows me to split the weight of the pack load between the two horses. To optimize return on each step, I try my best to place each foot straight in front of the other. My thoughts focus on finding the easiest route through the gantlet of cactus needles, rocks, rodent holes, and steep terrain.

Desert breathing comes deeply and slowly. Hot, still air makes it a task to breathe. Purposely, I slow my heart to maximize intake of oxygen with least effort, and minimize the loss of body water and energy.

Economy of motion becomes our living mantra.

Over several months, the horses and I have evolved to traveling as a herd. There no longer is any need for me to stay connected to them by a lead rope or with reins. Whichever direction

I walk, the horses follow. When I stop, the horses stop. I am the alpha of the herd. Or at least I like to think so.

As evening twilight's softening haze spreads its dark veil over the land, I begin to whistle a tune I am fond of. It is a tune with a rhythm that naturally keeps time with the percussive beat of the horses' hooves. Before long, I realize that I am whistling and walking alone and that the horses are no longer behind me.

What could have happened? My mind races: Did one of them fall into a rodent hole and break a leg? Did a rattler bite one of the horses? Did something cause them to panic?

I run through the desert night, retracing my trail with a flashlight as barbed needles pierce my pants and stab my skin. After a brief sprint, I am relieved to find Shalamar and Buck standing dead still. As I talk and touch them, Shalamar and Buck nuzzle my hands and chest. I look them over from head to foot—everything seems OK. My first inclination is that something spooked them. After catching my breath, I give Shalamar a few gentle tugs on her halter. She stands frozen in place as though anchored into the sand.

That's when I notice that Shalamar and Buck are keeping their eyes focused on a huge rock in the near distance, a prominent rock that juts skyward at a severe angle above boulder-strewn terrain about 50 yards distant.

Maybe a mountain lion is hiding in the rocks!

I draw the .22 magnum Ruger pistol and point the flashlight toward the jumble of rocks. I give a few loud shouts but hear or see nothing. I try again to have Shalamar and Buck follow me, but they remain rigidly in place while flicking their heads up and down in the direction of the towering rock.

Taking a deep breath, I stumble over the boulders toward the big rock with gun and flashlight pointing dead ahead. At closer inspection, I discover a deep cavernous depression hidden beneath the overhanging rock. When I shine the flashlight into the black void, my eyes are greeted by the magical reflection of sparkling light bouncing off the surface of water.

Crawling on hands and knees, I approach water's hiding place. My head bumps hard against the stone's wet belly. As I crawl up close, the clean, cool smell of dampness fills my nose. A few tiny frogs, each no larger than a horsefly, leap into the pool, piercing the surface and sending ripples throughout. Kneeling and bending as though praying, I see my undulating face looking back—an affirmation that I exist.

Slowly, I submerge my burning hands and give them a cleansing. With cupped, callused hands I gently lift the transparent life giver to my mouth. Stop. Evaporating vapors fill my nose with a fragrant bouquet of desert vintage. I say, "Thank you!" The words echo. I kiss water with cracked lips. Slowly fill my dry mouth with its wetness. Swish it around my tongue. Taste its exquisite essence as I fantasize: flavors of sand and rock, hints of blended root, stem and leaf, algae, and other things beyond belief—even the telltale blush of fish, frog, and pollywog! And what is this? I detect effervescing bubbly of bottom, gold and silver fermented by light of sun.

I am overcome by an ancient urge to merge. Sips deepen to gulps. A flood fills the bottle of my being.

I become light-headed and almost faint as the water surges through my body.

When little or no water is available to the body and one's survival is at stake, the brain's innate intelligence is to hoard water while rationing just enough to keep other organs functioning. When a person is dehydrated, the first gulps of water are absorbed sublingually, under the tongue, and injected into the brain. This is the body's natural survival mechanism imprinted since birth.

Every person will experience this phenomenon of having water rush to their heads in their journey. It may come from lack of water in the region of the world where you live, or after a long walk or jog, after shoveling snow, working out at the health club or playing sports, working around the home, forgetting to drink water while at work or during a long car drive, or just not being conscious of staying properly hydrated. Regardless, if you pay attention as you drink a glass of water when dehydrated, you may feel pain as the water flows down your throat into your stomach. Immediately, there will be an improvement in your mental and emotional well-being, including the loss of any headache or mental confusion that may have manifested during dehydration. After drinking a glass or two of water, you will find yourself refreshed, alert, and in a better mood.

It has been calculated that as one glass of water flows through your body's estimated 90,000 miles of blood capillaries, it will satisfy the needs of some 50 trillion cells. Also, once you are properly hydrated, your blood will carry up to four times more oxygen to your brain and body cells.

After drinking my fill and regaining consciousness, I hike back toward the horses. Along the way, I urinate for the first time in more than a day and feel sharp pain.

Fetching the collapsible canvas bucket from the pack, I make the trip over the rock field and return with a bucket of water for Shalamar and Buck. Upon smelling the water they fight viciously to see who will drink first. I am pushed backward, trip over a rock, and fall hard on the gravelly sand. The horses fight again as both try to lick the moisture from the empty bucket on the ground. It is a struggle to pull each of them away from the bucket. I grab some rope from the pack and tie them off to nearby rocks to create distance.

After the horses have drunk their fill, I give them a good wash-down to remove accumulated dirt and salts from their bodies, especially in the saddle areas on their backs; two things will endanger a horse's life during such a long trek—infected saddle sores and bad feet. Before hitting the trail, I learned equine first aid, including how to stitch wounds and how to trim and shoe horse feet.

Feeling tired from my many water-gathering trips over the boulder-strewn terrain, I sit to catch my breath. With a pair of Texas pliers I extract a few cactus needles out of my lower right leg.

Returning to the water, I crawl under the rock until the space gets tight. Strip naked. Turn off the flashlight, and enter the water slowly. With pounding heart I take a breath, immerse myself fully, and submit to water's loving embrace. For a brief second my body goes rigid, then relaxes.

I am reborn in the darkness of the watery womb. I feel safe, at home, and gradually the aches and pains of the trail dissolve—and my hot, sore, swollen hands become cool and soothed as the cold water triggers endorphins inside my body.

I relax and become one with water. Ears beneath the surface, I listen.

My tired, cuddled body drifts into sleep. When halfway between dreaming and waking, I hear a drop of water splash onto the surface from the rock above. It reverberates through the pool like a sustained musical note. A musical note composed by our universe billions of years ago. After a while, I hear the chirp of a frog. Then another. The reverberating musical tones of rhythmic water drops accompanied by singing frogs sound otherworldly.

I meditate on the water molecules created at the beginning of our universe. On interstellar water. On water vapor streaming off into space from the tail of a comet. On water's role in creating stars and cooling our sun. On water flowing through every life-form ever conceived. On water music. On thoughts and emotions born of water. On the rarity of liquid water throughout our universe.

A poem floats into mind:

Water
is
the ink
that writes
the poetry
of life

Throughout the starry night, I awake from time to time to fetch more water. Being hydrated, I better appreciate the shimmering beauty of the stars. I think about how all planets and life contain the stuff of stars. Stars born of water.

As I stare up into infinity, an occasional shooting star zips overhead. To me, the true wonder of meteors entering our atmosphere is that they can bring water. Even though there are several theories about the origins of Earth's water, there is no question that cosmic water from the universe did and still does make a contribution.

Recently, scientists using a radio wave telescope discovered water molecules from about 11.5 billion years ago. To think of ancient water molecules emitting electromagnetic waves that we can detect billions of years later with our modern radio telescopes boggles my mind. Such information confirms my belief that the vibrations of water molecules throughout the universe are interconnected across time and space.

NASA scientists have also recently learned that water usually serves as a cooling agent in the creation of stars. Even our sun has water in the form of superheated steam that keeps it from overheating.

As I up look at our night sky, it is reassuring to know that water exists wherever I see a star.

As the sun rises, our life-giving watering hole rapidly goes dry. Along the trail, I've heard tales of how desert springs and water holes will mysteriously flow to the surface only after sundown and then equally mysteriously disappear again after sunrise. Now I have seen it for myself. The water disappeared not from the amount we borrowed, nor from evaporation, but from the huge thirst of the surrounding vegetation sucking up the groundwater through their deep roots. In this fashion, the desert exhales and inhales water in harmony with the cosmic influences of our universe.

In some places around the globe, flowing springs are considered portals to the homes of our ancestors. Throughout human history, a succession of shrines, churches, and healing centers were located near such springs. These springs were said to have the ability to heal. Historical records indicate that special springs are frequently affiliated with a named spirit, often female.

As we make tracks away from the water hole, I look back at the magnificent jutting rock. I think of the ephemeral liquid elixir it hides. Had we passed the rock during daylight, the water would not have been present to save us.

And, based on the dry desert days to follow, I probably would not be here today writing this story.

While living outside during the two-year horseback trek—the biggest challenge was finding water. I never fell ill. I became ill only when I returned to living inside, where the cavelike darkness and unmoving stagnant air inside a house made me feel as if I were suffocating.

After the "Ride for Nature" adventure, I have continued my passion for learning from water. And I feel most blessed to be living a journey defined by water.

The embedding of water knowledge into my being while living outside for almost two years is most profound. From the horseback trek I learned that we cannot destroy water—but we can interfere with water's ability to support us and to create new life-forms.

Since getting out of the saddle, I've experienced water research adventures in more than 15 countries, including the Sahara and the Australian outback; founded and operated a state-certified water-testing laboratory for 14 years; founded and directed an environmental research institute for 20 years; published numerous books, poems, and articles about water; and spoken at the United Nations on World Water Day, and I was sponsored by the Council for a Parliament of World Religions to speak in Jerusalem before warring Middle East factions, in an effort to help them see water as a common denominator shared by all religions.

In my water travels and research over the decades, I have come to learn about our evolving knowledge of water. As a result, it has become clear that, as a result of water scarcity exacerbated by humankind, we will soon be forced to develop a new relationship with water. Just as we see trends concerning energy conservation, biodegradable packaging, recycling of wastes, and sustainable life choices, we will soon be adopting new behaviors relative to water.

On a broader note, the future health, wealth, and success of any country or civilization will be determined by access to adequate quantities of fresh water. Likewise—since water serves as a common denominator for all religions, political divisions, and life-forms—water will serve as the key that will help humanity find some semblance of peace.

When we bemoan water and air pollution, when we feel sadness at the loss of other species, when we feel concern over our children's future, we see the reflection of what we collectively have wrought upon this living world during our relatively brief tenure. Like flicking a light switch we can immediately alter the future by working with the creative energy of water.

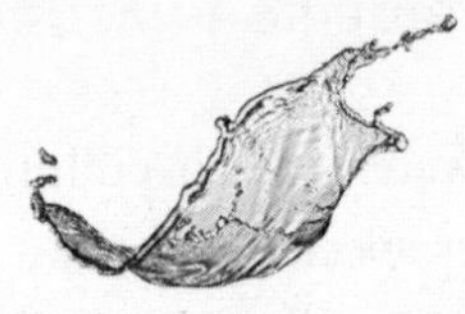

WATER CHANGES EVERYTHING

Scott Harrison

Scott Harrison spent ten years as a New York City party promoter, producing fashion and music events at top nightclubs. In 2004, disgusted with the indulgent and selfish life he led, he decided to volunteer aboard a hospital ship in Liberia, West Africa, as a photojournalist. Two years later, he returned to New York City to found a global nonprofit organization called charity: water. Turning his full attention to the world's 1.1 billion people without clean drinking water, he and his team created world-class exhibitions, innovative online giving campaigns, and nationally aired public service announcements. In less than three years, charity: water has raised massive awareness and more than ten million dollars, funding more than 1,390 water projects in 16 developing nations, providing over 700,000 people with clean, safe drinking water.

CHARITY: WATER was birthed in a SoHo, New York, apartment in 2006.

I was couch crashing and had written words describing global poverty issues all over the walls of my friend's place. In my solitary thought process, I had covered the wall with Post-it notes. I was 30 years old and on a journey toward embracing a newfound love for the poor after ten years of decadent living in New York City's nightlife scene.

Two years earlier, I'd ditched my world of $16 cocktails and $350 bottles of Grey Goose for the shores of Liberia, West Africa. There, as I was faced with extreme poverty in perhaps the world's poorest country, my life transformed while I was serving as a volunteer photojournalist alongside facial surgeons on an amazing hospital ship, the Mercy Ship *Anastasis.* I photographed thousands waiting outside stadiums for medical help in a country where one doctor served 50,000 citizens. I saw young children choking to death from giant tumors. I saw women my age blind from cataracts.

I simply couldn't walk away from what I saw. I was forever changed.

Upon my arrival back in New York, my destiny took a U-turn. I decided to throw the rest of my life into service for the poor. I knew that meant starting a charity—doing something that hadn't been done before. My friends in New York weren't giving enough money away. Many were disenchanted with the general concept of charity, citing bloated bureaucratic organizations and what they felt was a lack of transparency in the sector. In short, they told me they didn't know how much of the money they gave went to help the poor, and what impact it actually made. I wanted to start an organization that would bring people back to the table, engage them in a different way.

Full of faith, I conceived a new model. I'd find separate donors to fund operations and staff and prove to donors—using GPS technology, photographs, and video—that their money was supporting real projects that were actually being completed.

Next, I just needed an issue to launch. The word "water" screamed at me from the walls of that SoHo apartment. I'd seen people in Liberia drinking from swamps and rivers. The wells visiting volunteers built improved Liberians' health and brought dignity to their villages. With a billion people on the planet living without clean water, and 80 percent of all disease related to the lack of safe water and basic sanitation, this was an opportunity to make a difference. Water touches everything, it changes everything.

And that's how charity: water came to life, first with a few volunteers in a small apartment, and now, just three years later, with 15 amazing staff people in 8,000 square feet, managing 1,400 projects in 16 countries serving 700,000 people with clean water.

Over the last few years, I've traveled extensively throughout the developing world, and have had the chance to meet many of the people we've served. My ideas about water and poverty have evolved as I have listened to them and learned. I've been humbled many times, hearing about their daily struggles and courage—following them on their long journeys to dirty rivers, lakes, and swamps as they carry 80 pounds of water in yellow fuel cans, or dig with their children in sand for water, or line up at a well and wait eight hours for a turn.

The following are excerpts from field notes I wrote over the last three years. I invite you to get a glimpse of some of my travels and some of the work we've been able to accomplish.

NORTHERN UGANDA—GULU PROVINCE *August 4, 2006*

Stuffed in the back of a blue Toyota Land Cruiser with a human rights activist and two local Ugandan humanitarians, I hurtled north toward the Sudanese border. A military truck carrying five camouflaged army fighters with AK-47s followed closely behind, the soldiers choking in our red dust.

We passed several army roadblocks and convoys, necessary measures in Uganda's wild, wild northwest to defend travelers and Acholi villagers from Joseph Kony's dangerous Lord's Resistance Army rebel soldiers recently seen in the area. I took in a Ugandan sky with bigness hard to describe. It's a sky that seems to press too close to the ground as stacks of cumuli layer, then rupture to reveal patches of deep blue.

Blue feels out of place here.

Sixteen miles shy of the Sudan border, we reached our destination. The Atiak IDP camp was home to 23,638 internally displaced persons living in crude circular mud huts with thatched-straw roofs. They'd abandoned their villages ten years ago when their children were snatched from their homes by Kony's troops, forced to bear arms, forced to kill. More than 150 of the villagers were massacred in the town center in 1995. Now they find safety in numbers and safety in the camp.

"Take me to the water source," I asked Alex, one of the camp's leaders, who cheerfully led me outside the camp through tall grass and down a slope to a water hole. A few Acholi women were gathered there, drawing water from a source that some time ago would have shocked me.

The water was murky, stagnant, and unsafe. As Alex translated, I asked them why they were gathering water here, when there were ten borehole wells scattered throughout the vast camp. I listened and learned that the lines to take water from the cement wells were six to eight hours long. I was told that the wells were meant to serve more than 2,300 people each, but they didn't produce enough water. The women collecting murky water couldn't wait in line today, and like many others, had come to gather water from this fetid mud hole.

During my extensive tour of Rwanda last week, a little more than an hour south of its broken and busy capital, I saw men my age at a local source of drinking water. I choked back anger and tears as they gathered viscous, brown water from a swampy pit and poured it into filthy three-dollar jerry cans. Then they strapped the cans to bikes and hauled them miles uphill so the water could be used for drinking, cooking, and washing.

"Why don't they just boil the water?" I've asked more than a few times.

Rwanda's and Uganda's poor live on less than one dollar a day, so the answer is simple. They don't have the money to buy the charcoal that could be used to heat—and thus sanitize—their water. So they die because of the thirst that compels them to drink fetid water. The children die of diarrhea, as they drink contaminated water that dehydrates them and makes them thirst for more of the same. A vicious cycle of dirty fluid and death. They contract worms and parasites, and even nastier diseases with long clinical names and too many consonants.

ETHIOPIA *November 8, 2006*

About five hours south of Addis Ababa, the village of Bulgeta sits at 6,000 feet, and boasts almost that many inhabitants. To get there, our sturdy 4x4 snaked south through stunning mountains, and then turned right at a town called Shone. Thirty minutes later, we were navigating an impossibly rocky road full of children, cows, donkeys, and people pushing carts loaded with wood and water. We passed a primary school where hundreds were packed inside to learn reading, writing, and arithmetic, and moments later, we reached the source of water that had served the people of Bulgeta for generations—and stolen many of their lives.

For years, men, women, and children drank from the village's large pond. So did their livestock. Just beyond the pond lies a field where the cattle grazed. The sloping grass was littered with cow dung that, during the rains, was carried downhill to the pond. The water was filled with green algae and muck, making many in Bulgeta sick.

But as is the case in many poor communities in Africa, health care is out of reach for most there. Those who could afford treatment would be carried, on the shoulders of able-bodied relatives and friends, to the nearby Shone clinic, where they would receive antibiotics that cost between 40 and 80 cents. Those who couldn't afford treatment stayed at home and suffered. Many died from parasitic diseases, some from typhoid, and others from dehydration, often caused by plain old diarrhea.

The people of Bulgeta knew the water was killing them.

"It is not clean at all," Marcos, one of the village elders, told me in Amharic through a translator. "But we have no other solution. We have to drink it."

In my 11 days in Ethiopia, I visited many villages like Bulgeta. One of the organizations charity: water supports has drilled more than 150 wells in Ethiopia in the past few years, providing clean water to more than 750,000 people.

The drilling team comprises a remarkable group of expatriates and locals. The oldest, John Ed Clark, is a slim 69-year-old man with bushy eyebrows and furrowed brow. He told me that this was his 60th trip to the country. He has three more trips planned next year, and says plainly he'll keep coming until he can't.

I learned about the drilling process and even operated the controls of the 30-year-old Schramm rig with expert well driller Curt King. We looked for water on the grounds of a school for the deaf, and I was surprised to learn how complicated and technical the process was. A Seattle resident, Curt's been at this for almost three decades and reckons he's drilled more than 2,000 water wells in his lifetime. He's tall and kind and speaks shyly and softly about his work.

Trying not to get emotional, Curt talks of the women and children whom he really serves, the women who weep as he finishes a well. Women without a voice. Curt wants the work to go on when he's unable to continue, so he has been training three locals—Solomon, Nigusse, and Demoze—to take his place when he retires.

After leaving our drill team, I traveled another 1,000 miles over the next days to remote, water-stressed parts of the country and saw firsthand Ethiopia's great need for water. I photographed children digging in sand for the precious liquid. I saw hunched women with curved backs walk tortuous miles in the heat hauling heavy jugs of dirty fluid.

Touring a health clinic that served 103,000 people and had no doctors, I watched the administrator flip through the patient logbook, seemingly shocked himself at the incidence of waterborne disease. His clinic saw about 50 percent, which, although not quite as high as the often quoted UN number of 80 percent, was alarming. When I asked him to tell me more, the administrator jumped in our truck and drove us to the local source he said was responsible for the illness. We watched cattle and children share the same drink from large, brown ponds—a scene that now was familiar to me.

Women washed clothes, bathed, and drank in the same place. We stood over a small, stagnant hole and watched as a young boy with swollen feet, wearing a ragged, gray sweatshirt, gathered two and a half gallons.

MONROVIA, LIBERIA *March 20, 2007*

I spent the last seven days seeing about the "water business," as they say here in Liberia. Travels with my camera took me through the slums of Monrovia to dusty towns in the jungle, where I followed children as they fetched water from fetid ponds and swamps. I met some incredibly resilient people, more faces to the world's 1.1 billion without access to clean water.

In these places, most are fortunate enough to live in a small house with a good roof of zinc and splintered wooden windows that swing open on hinges. The conditions, however, are vile. The dry and dusty area around your house is littered with garbage and bile. There's no sewage system in town—or in the country, for that matter. The public latrines are ramshackle tin structures built over the river, on the shores of which squat small children. Human waste floats downstream, and at the beaches, the golden sand reeks and the ocean waves carry fecal matter. You dread the rainy season's approach in late May. The downpours will come, the rivers and swamps will rise and sweep, and your neighborhood will again turn into an open sewer.

But today is dry and hot, and you need water. You have a few options.

The first is to buy water. You've got some money in your pocket. You grab your two five-gallon jerry cans and head over to a nearby city "tap" for water. There's a wait, but you stand patiently in line for 30 minutes and pay the attendant ten Liberian dollars (about eight cents) for each jerry can. He grabs the dirty vacuum cleaner hose attached to the city tap, and fills your buckets. There's a square cement hole next to you that collects the spillage, and it's a nasty mess you avoid stepping in.

The second option is the open well. You had a bad day selling your wares at market yesterday and don't have a single Liberian dollar to your name. You hope business will be better today, so you can at least get something to eat. But you need water now. Instead of the city tap, you head to one of the open wells nearby. You use a shredded tire to lower a badly broken algae-coated bucket into the hole and lift the water out. The water comes out milky and white, so you head back to your back porch, where you've got a bag of sand hung from sticks. You dump the well water into your homemade filter and place a bucket underneath. Sand is amazing at filtering out contaminants, but you couldn't afford to buy clean sand, and instead got this sand from the beach, which also serves as Monrovia's toilet for thousands. The water looks clean as it comes out, and you hope for the best as you drink, cook, and clean with the ten gallons you collected.

There are no good options if you happen to live in the countryside near Liberia's border with Guinea. Your surroundings are lush, green, and beautiful, and nights are cooler than in the city. Malaria is a killer here, but you've had it so many times now, your body has worked up some resistance to it, so that it's more of a regular nuisance than a threat to your life now. You live in a mud house with a leaky roof in a town with about 750 others, and you'll be lucky to make $150 this year. You farm a root called cassava, and can generally keep food on the table. You use most of the 50 cents you make each day to buy rice, fish, and soap.

Water is big trouble for you here.

Unlike in the city, with its conveniences, here you've got to walk quite a distance into the jungle to find the pond where you'll gather your water. The village pigs use the same pond to bathe and defecate, and even though the water looks pretty clear, you hate the fact that everybody else from the village wades into the middle to collect water. You've had diarrhea for three weeks now, and suspect the water. You would carry yourself to the nearest clinic 15 miles away if you had the money for transport, but even if you got yourself there, the two-dollar admission fee and three-dollar medication would be out of reach. You suffer silently and hope your stomach will return to normal soon.

ORISSA, INDIA *January 1, 2008*

I never thought I'd be spending New Year's Eve looking at toilets. And I never thought I'd be asking my friends for money to build them.

Like many things in my comfortable Western life, toilets have always just been there for me. They were usually made of white porcelain, held about five gallons of clean water in their tank, and were in rooms with doors I could lock. Yet 2.5 billion people on our planet don't know what any of that is like.

"Cover your face and expose your base," Joe Madiath says darkly with a chuckle.

It's New Year's Eve and I'm in rural East India, talking about—what else—toilets and open defecation. Orissa is India's poorest state, and the 150-acre compound that we are at is where Joe lives and works. It's truly in the sticks—about four hours from the nearest city.

"It's all about dignity," Joe says. I nod my head fervently and agree, trying to process what I've just seen.

Joe is preaching about the right to basic sanitation that one-third of the people in the world currently go without. In layman's terms, this means not having a toilet to use. Here in Orissa, squatting in open fields or forests is a daily reality for 99 percent of the people. A staggering 94 percent also don't have clean, safe water to drink. These injustices—or indignities, as Joe calls them—have for more than 25 years motivated his humanitarian work with the organization he founded called Gram Vikas.

Joe's a gruff, bearded man, about five feet seven, who walks and talks with authority and purpose. He's led a fascinating life, both as a student activist and then as a champion of human dignity for Orissa's rural poor.

He started young. When he was 11, he thought the workers at his father's rubber plantation should stand up for their rights; he got them together and helped them form a union. His dad shipped Joe off to boarding school, and he never returned home.

He lives modestly with his wife, Shirley, on the edge of the compound in a house that he says nobody else wants to live in. After a few meals there, I saw what he meant. Shirley joked that the house—sparsely decorated with Van Gogh posters and Indian artifacts—might fall down any moment.

My first conversation a few weeks ago had left me eager to meet Joe. I'd called his cell phone from New York to apologize in advance for making him take us to see field projects on New Year's Eve. It was as good a day as any for me to work, but I wanted to be respectful of his holiday celebrations.

"If you fly all the way over here because you want to help our people, the least I can do is show you around," he said. "Every day is a work day for us here."

And show us around he did.

Ten and a half hours ahead of the year-end celebrations back home in New York City, we spent the holiday visiting rural villages in Orissa. Some were so poor that entire families of five lived on only 50 cents a day—$150 a year.

Gram Vikas does many different things to lift people out of extreme poverty, and Joe doesn't believe that just because people are poor they should have poor-quality solutions. Almost all Gram Vikas work begins with clean water and basic sanitation. So few of the communities here have access to clean and safe drinking water, and many of the people who do have it walk miles for that water.

For example, before Gram Vikas helped the village of Khatuakuda get clean, piped-in water, Manu used to wake up at 3 a.m. and spend four hours each morning fetching water. She'd then spend two hours in the evening doing the same. Imagine, six hours every day to fetch water.

Joe's solution to bring water to these rural villages is impressive and cost-effective. Gram Vikas will work in a village only if 100 percent of the community "buys in" to the work. The caste system makes this interesting, and some villages won't let the *dalits,* or "Untouchables," anywhere near their water source.

Some communities, eager for solutions, are breaking old molds: Joe introduced me to an "Untouchable" who was actually elected head of the water committee by the village.

The 100 percent buy-in has taken some villages as long as a decade, but most often only a few months, as many desperately want Joe's help. The help comes at a price, as community members must subsidize development costs, and provide the labor for the project.

HONDURAS *July 28, 2008*

The Río Plátano starts high in the Honduran Mountains of La Mosquitia. After hundreds of switchbacks, it ends 70 miles later in a town that shares its name, throwing a massive delta of brown sediment into otherwise turquoise coastal waters.

Río Plátano town is a dive, but not the kind of dive one might associate with Honduran tourism or exotic islands. There are no gringos with dollar bills to be found here, but instead a small, forgotten population of about 400 people nestled between the ocean, the river, and a marsh. There's very little clean water, few decent toilets, no soap, and not much hope. The homes are built on stilts, and the people are slowly drowning in the high water table—dig down 12 inches anywhere in town, and you get soggy. Unsafe water covers everything.

DRINK. If you'd like a drink in Río Plátano, you'd choose your poison from a series of toxic holes in the ground. Some are boxed neatly with wood from the forest; others are open, their owners not bothering to protect the murky green surface water. A few of the houses direct bits of tin roof into plastic barrels to catch rainwater, a decent solution only during the wet season.

When the fish are plentiful, incomes are higher and some of the people can afford chlorine to purify their water. Most people don't use chlorine, though, and so they regularly suffer with common waterborne diseases—diarrhea, parasites, skin rashes, and bouts of vomiting.

FLUSH. If you wanted to use the toilet here, you'd head away from the beach and toward the marsh. Then up a narrow plank to a small wooden shack on stilts and through a crooked door. But instead of white porcelain and a handle to flush, you'd find an unpleasant hole that drops waste ten feet below into open, marshy ground. The stench would overwhelm you.

WASH. If you wanted to wash your hands, you'd be out of luck. The local store doesn't have soap in stock, and it does not seem to expect another shipment anytime soon. And the one-dollar buy-in is too steep a price for many who are just struggling to put food on the table. Forget the clinic if you're sick. A woman sits here with a huge box of unopened medication and an attitude of defeat. Although she might see 10 to 15 people a day, few leave better off than when they came.

The elementary school is Río Plátano's lone bright spot—at this place, four teachers fight despair with education. The headmaster is 34-year-old Denuer Idin, and he's big on hygiene even in a town without soap. He and his crew inspect the hands of the 91 students and send kids showing dirty hands home to scrub them clean. The school has managed to get cement-block latrines built that are closed and kept reasonably tidy. They are color-coded white and blue for boys and girls. Denuer says bad water keeps kids out of school for a variety of reasons, but he'd have more kids studying if clean water were available.

BACK TO ORISSA, INDIA *Thursday, January 8, 2009*

I jumped back but too late, as coconut juice splashed all over my jeans. The upside-down transfer from coconut to steel cup hadn't factored in the volume differential.

I sat on a plastic chair in the Engreda village Baptist church, tucked away in the rural hills of eastern India. Men and women of the community had gathered here to thank us for funding a piped water system that brought clean and safe drinking water down from a new well in the mountains.

Junash, 41, was the one who had spilled on me, but I didn't mind, and I drank two cups of the warm juice. A few minutes earlier, he'd made a speech about what happened here. I learned that the 567 residents of Engreda had big problems with water. Their primary source for years had been a polluted stream in the valley beneath the village, which I saw a few moments later.

"In the stream, we would remove a little bit of sand, and the water would ooze out into it. We used to drink that, and the children and adults used to get diarrhea," Junash said. "We are poor. Whatever savings we had, we spent on curing our waterborne diseases. The poor remained poor."

Not anymore.

charity: water had funded a well at Engreda, but the water running from their taps came at a price higher than our funding. The people had petitioned our implementing partner Gram Vikas to help them with the water problem. But before bringing clean water to Engreda, Gram Vikas asked villagers to give a year of their time to construct toilets and bathing rooms on faith.

For more than 30 years, Gram Vikas has taken a unique approach to development work. For it, sanitation is the key to good health, and community participation is the key to sustainability. "Sanitation" meant toilets and showers here; "participation"—a year plus of hard work.

Junash said Gram Vikas's proposal was initially met with some resistance, as each of Engreda's 130 families would have to do a "lot of work" that would cost "a lot of money." For Gram Vikas to work in a community, 100 percent of the people must agree and contribute, and after a short time, they did.

But its involvement didn't stop there.

After all 130 toilets and bathing rooms were constructed, community members then helped lay pipe from the well Gram Vikas constructed high in the mountain near a spring. It was tough going. Villagers spent more than a month breaking stones in the rocky ground, beaming with pride at their achievement.

Written on the wall next to our contribution was theirs, and although not in the form of a check, its value far exceeded ours.

The stone, bricks, gravel, and labor the people of Engreda added to the project came to $19,851. At least half of that was sweat equity and calculated at the going rate of 17 cents an hour. For comparison, if their labor took place in the United States, where hourly minimum wage is $6.55, they'd have contributed more than $380,000 of labor value—58,345 hours. In that light, charity: water's $7,822 contribution for the hard costs of piping, taps, and the water tower was a steal.

Back at the Gram Vikas compound later that evening, the project coordinator smiled when she learned I'd visited Engreda.

"Yes, they're very happy there. They tell us the water tastes better than coconut milk."

I had to agree.

SAVE WHAT'S LEFT

Lynne Cherry

Lynne Cherry is the author and illustrator of more than 30 award-winning children's books, many that teach children respect for the Earth. Her best sellers *The Great Kapok Tree* and *A River Ran Wild* have sold more than a million copies and been translated into many languages. Cherry has produced four short movies. *Young Voices on Climate Change,* which was screened at the American Museum of Natural History, tells the stories of young people who have helped to reduce the carbon footprints of their homes, schools, communities, and states. Cherry lectures widely and passionately to school groups and other organizations about how kids can make positive changes. She also speaks to educators about how to integrate nature studies into their curricula.

I LIVE AT THE TOP OF A WATERSHED on a mountain farm, and I often think: This is the way the whole world used to be. From the upstairs window of this Civil War log house I look out upon a landscape of bygone days—golden hay fields crisscrossed by swooping barn swallows, a hundred-year-old apple orchard with resident bluebirds and a forest of majestic hickory and tulip poplars all under a vast ever changing sky. A cobalt blue canvas can suddenly transform as a gray velvet cloak of clouds extends its fingers eastward, frothing and billowing, flashing and rumbling—the precursor to an imminent torrent.

The hissing rain descends in a sparkling curtain, and often these prolonged summer showers permeate the ground beyond its capacity to absorb. Saturated fields can hold no more water, so springs bubble up from the ground, seeping out in a dozen different places from any crevice they can find. One, erupting like a miniature Old Faithful, springs forth from a foot-deep hole under a blackberry thicket. To the birds, and untold numbers of other living beings inhabiting this thorny tangle, this water is lifeblood. And here a watershed is born.

This is the water cycle come to life—water vapor carried by clouds, cooling as they pass over the mountains, dropping their moisture as rain that saturates the land, creating seeps and springs that flow into streams and ponds and wetlands, continuing to flow down and down to the river, and into the ocean. Ocean water, heated by the sun, evaporates, forms clouds, is carried by the wind over the land, runs into the mountains and drops its moisture, and the cycle begins anew.

By the thicket stands an old springhouse. The first time anyone takes a drink from the spring sheltered here their eyes widen and they exclaim, "This water is *good.*" This water *is* good. Actually, it's beyond good. It's transcendent, transformative, restorative, remarkable. It has the flavor and the fragrance of the forest. Once upon a time all water used to taste this good.

Several other springs emerge from beneath great moss-covered boulders and merge to form rivulets, rushing downhill, downstream, pulled by gravity, swelling the stream that flows into the swimming pond, which rises, overflows, and cascades over its banks into a wetland bog spreading over the forest floor. Here the water can stand and wait until the ground is able to absorb it, at which point it percolates deep, purified on its descent, and recharges the underground aquifer.

This bog reminds me of one of my childhood haunts—a swampy place on the way to the "crick" where I spent every summer day, sunrise to sunset. Once, after a rain, the clay-colored ground began to undulate around my yellow boots. Kneeling down for a closer look I discovered hundreds of minute clay-colored frogs, smaller than the fingernail of that child I was long ago, and they were springing around like jumping beans, exulting in the rain.

But one day I came home from school to find my world destroyed. A bulldozer was erasing my forest, my stream, the bog—with all its inhabitants—from the face of the Earth. The deep sadness of losing the one thing that was the most important to me—the place that, in essence, defined me—followed me throughout my life as profound loss often does.

In my young psyche I realized that things were not as they should be and that perhaps these adults did not really know what they were doing. Certainly, when they destroyed my forest, no adult had the intimate knowledge I did of what lived there, nor did they know or care how profoundly they were affecting my life and those of the forest creatures. I believed someone should tell them. Like the child in *The Emperor's New Clothes,* whose shouting out the obvious takes the blinders off the eyes of adults in denial, I dreamed I might someday help the grown-ups to see the world with new eyes.

As an adult, when I first learned about the destruction of the tropical rain forest, those feelings of loss buried so long ago bubbled to the surface. It grieved me to imagine, with the rain forest's astonishing biodiversity, how many living beings were losing their habitat and their life. I thought, I need to write a book to let kids know what's happening. Maybe they can do something to save these sacred places.

I wrote that book, which would eventually be titled *The Great Kapok Tree.* The story goes like this: A man ventures into the Brazilian forest to cut down a huge tree, but before he can do so, the heat and the hum of the forest lull him to sleep. As he sleeps, a panoply of rare and wondrous creatures come and whisper in his ear a litany of reasons why he should not cut the tree. "*Senhor,*" says the sloth in her deep and lazy voice, "if you destroy the beauty of the rain forest, on what would you feast your eyes?" A native child whispers, "Senhor, when you awake, please look upon us all with new eyes." The story reaches children's hearts and minds while conveying lessons in empathy and ecology.

The text was finished, but the book still needed illustrations, so with my manuscript in hand I traveled to the Amazon rain forest in Brazil to research trees. I learned that mahogany trees were plantation trees—not the best representative of the rain forest. The biologists I met all agreed: I should feature the kapok tree, the quintessential rain forest tree. The kapok is the tallest tree in the forest, known as an emergent because it towers higher than the other trees of the forest canopy, and when in bloom is visible for miles around, with its profusion of yellow flowers. It is also known as the silk cotton tree for its seedpods, which burst open to reveal the softest—and most buoyant—substance imaginable. This used to be what flotation jackets were stuffed with during World War II. The kapok tree, with its dramatic buttressed roots, home to myriad creatures, was clearly the best centerpiece for this tale of the Amazon rain forest.

This was a place defined by water. It rained every afternoon, when the sunny sky gave way to a tropical deluge. I felt like I was at home there. The shade from the forest helped retain the moisture in the forest floor, and the aroma of the rain forest brought me back to my childhood. The ratio of my height to the height of the canopy was the same ratio as my height when I was a child to the top of my long-ago forest, so I again felt very, very small under these vaulted arches.

Subsequently, I was asked to teach courses in rain forests around the world, including in Peru. I saw myself as a translator—translating the language of science and making it accessible to children, and hence to the general public. Year after year I returned to the Peruvian rain forest to teach rain forest ecology to elementary through college teachers. Standing at the top of the canopy walkway, 50 miles from Iquitos, I could see no evidence of the human hand upon the land, just unbroken rain forest from horizon to horizon in every direction. The rain forest itself was my greatest teacher, reminding me of my relationship to the Earth in a way that I had not experienced since childhood.

Visiting rain forests around the world has allowed me to relive those halcyon days. For two weeks I had the privilege of living with the Tirio Indians, a "primitive" Amazon tribe in Suriname where ethnobotanist Mark Plotkin had been living for more than 20 years. Mark and I wrote *The Shaman's Apprentice* in the village of Kwamalasamutu, sitting under a thatched roof, with tarantulas intermittently raining down upon us. In this simple hut with a dirt floor and no electricity and, each night, billions and billions of stars overhead, I thought: This is how all humanity used to live—simply, and connected to all the treasures of heaven and earth.

The village of Kwamalasamutu is situated along the Sipaliwini River. There is no road; the *river* is the road. We launched our dugout canoes and traveled along it. Sitting on the riverbank, listening to the haunting nocturnal calls of the nightjars, I experienced total darkness. As my retinas expanded to take in every small bit of light, the universe seemed to expand. I could see deeper and deeper as faraway stars became visible. Shooting stars left stardust across the inky sky.

These forest people shared their knowledge of their natural world. The shaman taught Mark about indigenous plants that have great healing power, which they believed to be more powerful than modern medicine. Because their life depended on it, the people in Kwamalasamutu were also very attuned to water, weather, and climate. Changes in weather patterns could necessitate their migrating to another place to sustain their existence.

Photojournalist Gary Braasch and I visited the rain forest encampment of biologist Steve Williams in the Daintree wet tropical forest in Australia to research and write our kids' book on climate change. For decades, Gary had been on assignment in rain forests throughout the world—and also in the Arctic and Antarctic, where he witnessed firsthand the dramatic changes in ice cover and sounded the alarm with his book *Earth Under Fire*. Dr. Williams told us of his findings: As the climate changed, the animals are moving up the mountains to stay within their climate comfort zone. Frogs, for example, that needed the water vapor of the cloud forests to survive had to follow the clouds or die as the clouds relocated up the mountain. And for frogs already living on the *top* of the mountain, there was no more "up."

In rain forests throughout the world I gained a deeper understanding of how, in protecting species and protecting forests, we are also protecting our water supply and the carbon sinks that can mitigate global warming. I saw how water conservation, habitat conservation, and global warming are integrally interconnected.

The mornings in the Australian rain forest were the most otherworldly of any place I've ever visited. Never, anywhere else in the world, have I awoken to the cacophony of so many different birdcalls all singing in a jumble like a Philip Glass symphony. And the splendid flashes of color accompanying that joyful noise were splashes of pure wildness.

Returning from the rain forest's smells and simplicity, and the sense of having traveled back to a more human-scale place and time, reentry into the "civilized" world was quite shocking. It does indeed induce culture shock to see how we have trod so heavily upon the Earth in our mechanized world.

While speaking on behalf of a conservation effort, I appeared on a TV show in Florida, where I told my personal story about losing the land that I loved. Charles DeVeney, a physical education teacher in Coral Springs, saw the show and contacted me to share his story of losing the childhood places he'd cherished.

He had lived in Coral Springs all his life, and he loved exploring its swamps and springs. There were springs that came up from underground caves that you could actually dive into. He became a teacher, and every day he rode his bike to school. Although most of the area had been developed, there were a few places that remained the same as they had been during his boyhood.

One morning, as he approached his favorite cypress swamp, he heard the hum of chain saws, and he watched in horror as the tall trees came crashing down. He was choking back tears when he arrived in his homeroom class. "What's wrong, Mr. DeVeney?" asked one of his students.

Realizing that something was indeed very wrong, the students sat and listened, transfixed, as he told them how, as a boy, he could spend summer days exploring the wildest places, glades and marshes full of wading birds and terrapins. But little by little, year by year, throughout his lifetime place after place was replaced with shopping centers and housing developments. He told them how much he had loved the forest that had been cut down that day and how much, every day since he began teaching at Coral Springs High School, he had looked forward to momentarily leaving the asphalt and coming to that green place on his early morning bike ride. "At this rate, there will be nothing left by the time you're my age," he told the kids, despondently.

The students asked Mr. DeVeney if he could think of any way that they could save the swamp. He told them the ways he knew to save open space—through purchase by a conservation group like the Nature Conservancy or the Trust for Public Land, or through the action of citizens to raise the money through a bond that would be voted on in an election. The students decided to try to collect enough signatures to get a bond issue on the ballot in the next election to save the last remaining open spaces in Coral Springs. They called their group Save What's Left.

When the citizens became aware of the kids' campaign, they came from far and wide to sign their petition. On voting day, more people turned out to vote than ever before in the history of Coral Springs, and the ballot initiative passed overwhelmingly, saving the cypress swamp and the other last undeveloped places.

While collecting signatures, the students realized that most Coral Springs citizens did not understand the connection between wetlands, open space, and ensuring their water supply. They also saw a lot of evidence of ecophobia—people having negative images of swamps, fearing them as mosquito-infested, dangerous places. The students decided that they needed to educate Coral Springs citizens and allow them to experience the cypress swamp's mysterious dangling Spanish moss, the sinuous limbs of giant oak trees and magnificent long-legged wading birds. The kids wanted the community to see, feel, and come to really *know* the amazing place that they had saved. They invited local business people and others from their community to come visit their magnificent swamp; they taught them about its value and raised funds for a walkway through the big cypresses.

Save What's Left has inspired me to do what I can to also save what's left. I am placing my own farm under conservation easement and can rest in peace knowing that a part of a bygone era will persist on the top of Catoctin Mountain. Others before me protected 10,000 acres of parkland bordering the farm, ensuring the quality of the water that originates and flows through here.

Water is in my blood. Sometimes I think I must have been a fish in another life—or an otter, reveling in the feeling of gliding through a river. I have spent many weeks exploring rivers from their headwaters to their mouths on Sojourns—the Susquehanna River Sojourn, the Potomac River Sojourn, the Patuxent, Monocacy and James River Sojourns. As we paddle downriver in our kayaks and canoes we sing this ancient chant in rounds:

The river is flow-o-ing
It's flowing and grow-o-ing
The river is flow-o-ing
Down to the sea.
Mother, carry me
Your child I will always be
Mother carry me
Down to the sea.

Just as I lost my beloved forest—my childhood world—millions of people in today's world are losing their water, their land, their lives, and *their* worlds. As a young child I did not know that I had any power. But kids today know that they do.

As with any great movement, first, like a few raindrops falling, just a few young people speak out. They inspire others, and a trickle of them work to save land, protect their water resources, save rain forest, save what's left. The trickles come together into a stream until a great river of thoughts, of actions, of great expectations flows. Flowing and growing out to the sea, an ocean of humanity will change our collective destiny, chart a new course, and change our relationship to the natural world. That is my hope.

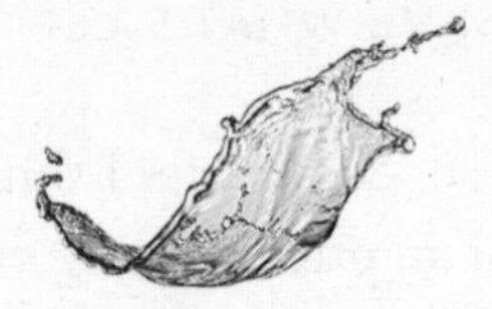

TO SAVE A GULL

Alex Matthiessen

Alex Matthiessen graduated from the University of California at Santa Cruz in 1988 with a B.A. in biology and environmental studies, and in 1995 he earned his master's of public administration from the John F. Kennedy School of Government at Harvard University. His work for the White House Council on Environmental Quality, as well as for the Harvard Institute for International Development, led to his appointment as a special assistant to the U.S. Department of the Interior in 1997. There, he was co-creator and head of the Green Energy Parks initiative—a joint program between the National Park Service and the Department of Energy—for which he received a Presidential Award from the White House. In 2000, Matthiessen became chief executive of Riverkeeper, the environmental nonprofit that protects and defends the Hudson River and the New York City watershed.

ONE SPRING DAY, when I was seven or eight, my father and I went on a bicycle ride near our house in Sagaponack, New York, on the east end of Long Island. On the way back, we came across a seagull hopping around on one leg. It had tangled itself in a discarded plastic six-pack ring. This type of plastic-ring packaging, used to tether together cans of beer or soda, has long since been banned because of the harm it caused to wildlife. This gull had managed to insert its head through one of the rings and a foot through another. With its leg in a "sling," it might as well have had a black eye, an Ace bandage around its head, and a crutch under the opposite wing. It looked almost comical, except it was really upsetting.

My father and I dismounted our bikes and approached the gull slowly, hoping to catch and free it. But of course, every time we got close, it flapped its wings just enough to get a few feet airborne and land at a safe remove from us. After a time, we gave up. Walking back to the bikes, Dad told me the gull would probably die because it wouldn't be able to feed itself.

As we pedaled home, I cried.

I felt a sense of helplessness that I projected onto the bird. Here it was, alone in a potato field next to a suburban house, far from the seashore, apparently abandoned by its flock to fend for itself. As a little boy who had just lost his mother to cancer, I too felt vulnerable and alone, an innocent victim of an adult world of death and dying that I didn't understand. At that point, and for many years afterward, I had not shed a tear for my mother; mourning the death of that bird was somehow easier.

The other emotion that stayed with me into adulthood was a feeling of shame caused by the injustice of the bird's unnatural death. Witnessing that bird's struggle gave me a feeling of profound sorrow, not only because it was going to die, but also because I understood that we humans were responsible. I suppose it was on that day that I became an environmentalist.

My father, a naturalist, author, and environmental advocate, was an early and important influence on me. Because of him, I became interested in birds and wildlife (my "bird name" is the Dunlin). He introduced me to the idea that human "progress" has a dark side. His book *Wildlife in America,* which came out just a few years before Rachel Carson's *Silent Spring,* is an elegiac chronicle of our subjugation and destruction of nature since settling the country. It is an unsettling description of what we've lost.

As a child, I was lucky to spend a few summers in East Africa, where my father was working on several books about the people and wildlife of the region. My stepgrandfather, a German doctor who had fled to Africa after escaping the Nazis with his young Jewish wife, ran a hospital in Njombe, in the southern highlands of Tanzania. Eck as he was called, would load us up in his sage green British Land Rover and take us on safari. To this day, nothing has been more thrilling than being among the big-game animals of the Serengeti, as the plains of northwestern Tanzania are called.

I rediscovered my passion for wildlife in college. I spent my summers backpacking in Alaska and the western United States and a semester in Montana's Glacier National Park studying grizzlies, bald eagles, and gray wolves. After college I thought I wanted to be a wildlife biologist. I took a volunteer job working for the Wildlife Conservation Society in the Ituri rain forest of northeastern Zaire (now the Democratic Republic of the Congo). For six months, I lived deep in the forest, in mud and thatched-roof huts, and worked alongside Pygmy and Bantu peoples. Roaming the forest among black mambas, Gabon vipers, and short-tempered forest buffalo, we tracked and collected data on the okapi, a gentle forest-dwelling "cousin" of the giraffe. Realizing I didn't have the temperament or patience of a scientist, I returned to the United States and worked for various nonprofit environmental groups before heading to graduate school to study public policy, which I believed would put me in a better position to help bring about more systemic improvements in the way we interact with nature and manage the Earth's natural resources.

In 1997, I joined the Clinton Administration as a political appointee to the Department of the Interior, in Washington, D.C. Toward the end of President Bill Clinton's second term, a mutual friend put me in touch with environmentalist Robert F. Kennedy, Jr., who came to Washington to encourage me to apply for the top job at Riverkeeper, a Hudson Valley–based environmental organization. At the end of meeting, he said to me, "You seem to suffer from the same angst about what we're doing to the planet as I do. And I can see you're not going to stop trying to do something about it until we succeed. That's why you're perfect for Riverkeeper." In 2000 I joined Riverkeeper and became a water advocate.

One night in 1966, a group of fishermen gathered in the American Legion Hall in Crotonville, New York, to voice their outrage over the polluted state of the Hudson. The river had been treated as an open dump for chemical and sewage waste as far back as anyone could remember. By the mid-1960s, it had become so polluted that parents forbade their children from dipping a toe in the water, let alone going for a swim or catching a fish to bring home for supper. Designated on American Geographical Society maps as an "industrial waste conveyance," the Hudson had become the butt of jokes on late-night television. But, for the commercial fishermen in the room, it was serious business, as their livelihoods were at stake. Angry and riled up, they started plotting to sabotage the waterfront facilities that were discharging the waste.

A *Sports Illustrated* writer and angler named Robert H. Boyle who was there that night offered an alternative strategy: Rather than break the law, they should enforce it. Boyle had unearthed two little-known 19th-century statutes that provided a bounty to anyone who turned in evidence of pollution that led to a conviction. Calling themselves the Hudson River Fishermen's Association, the group determined to track down and stop every polluter on the Hudson. They turned in evidence against Penn Central, a railway company that was discharging oil directly into the river, and collected a $2,000 reward. Soon after, they joined Scenic

Hudson in the legendary battle to save Storm King Mountain, in the majestic Hudson Highlands, which Con Edison had hoped to turn into a massive hydroelectric facility. Thanks to precedents established in the Storm King case, those 19th-century statutes were supplanted with more modern laws like the federal Clean Water Act, which for the first time gave citizens the right to bring lawsuits against polluters directly.

In 1983, Boyle and "the Fishermen" hired an activist named John Cronin to be the organization's first "riverkeeper." The concept of having a full-time sentinel stationed on the river patrolling for polluters was so powerful that the group soon adopted Riverkeeper as its name.

Having a patrol boat out on the river is critical to our work. John Lipscomb, Riverkeeper's patrol boat captain, is a passionate, highly knowledgeable advocate and ambassador for the Hudson; everyone who works or travels the river knows him and the Riverkeeper boat. John is out on the river ten months of the year, six days a week, as many as sixteen hours a day, looking for signs of illegal discharges and acting as a deterrent to would-be polluters. In a typical response to his presence, a marina owner once said to him, "You keep guys like me honest." Not only does it remind polluters we're watching them, but it inspires and comforts people who live and recreate on the Hudson to know there is a group dedicated to defending it.

A story that particularly touched me involved a little girl of about ten whose family was traveling up the Hudson in a luxury yacht that was docked behind our patrol boat. After a few conversations with the girl's father, it had become clear to John that he was generally hostile to environmentalists—a hostility captured by the copy of our newsletter that appeared to have been unceremoniously tossed into the garbage can on the dock not long after

John had given it to him. But, despite her dad's apparent inclinations, this little girl had gotten the message and was appreciative of our efforts. The morning her family departed, she left, along with a mussel shell that John keeps on the dash of the boat's cockpit to this day, a note that read: "Dear Riverkeepers, Thank you for helping out with the river. I'm glad you're helping nature. It feels so good to help. Have a safe trip, Sincerely, Colleen (the boat in back of you)."

More recently, a college-age intern named Rob told me how he got involved with Riverkeeper. He recounted coming into our old office in Garrison, New York, with his parents to pick up some Riverkeeper baseball caps he was going to give to his friends at his bar mitzvah. He reminded me that, in meeting him, I had asked what he wanted to do when he grew up. When he shyly admitted that he didn't know, I suggested he find a river in trouble and become a riverkeeper. Now in college, he has been an intern with us for three summers in a row and still aims to become a riverkeeper.

In 1987, Riverkeeper launched a campaign to safeguard the upstate watersheds that provide drinking water to half of New York State's 20 million people, including those who live in New York City, Westchester County, and a half dozen other Hudson Valley towns. The New York City Watershed, as it is known, actually consists of three watersheds, nineteen reservoirs, and three controlled lakes covering 2,000 square miles east and west of the Hudson River, making it the largest unfiltered surface water supply in the United States. Protecting this system, most of which is located within the Hudson River watershed, quickly became a central part of Riverkeeper's mission.

By 1989, the U.S. Environmental Protection Agency (EPA) was threatening to slap New York City with an order to build an expensive filtration plant to treat the water coming from the Catskill/Delaware Watersheds, which originate west of the Hudson in the Catskill Mountains and supply 90 percent of the city's water. The city had already agreed to build a filtration plant for the Croton system, which, located east of the Hudson in the much more populated and developed Putnam and Westchester Counties, was already showing signs of irreversible pollution.

Riverkeeper's chief prosecuting attorney, Robert F. Kennedy, Jr., launched a crusade to persuade the city to avoid the EPA filtration order by investing in protecting the Cat/Del water at its source, a far less expensive option than filtration. To make his point, Kennedy found a local real estate agent who calculated that even if the city bought every acre of land in the Cat/Del Watersheds to protect the water supply, it would cost only one billion dollars (in 1989 dollars), compared with an estimated six billion to eight billion dollars to build the filtration plant, plus another $300 million to $400 million to operate and maintain it each year. The comparison provided a dramatic example of how it is always more costly to treat pollution than to avoid polluting in the first place.

Throughout the early to mid 1990s, as a way to keep the pressure on city officials, Riverkeeper ran a sustained media and ad campaign designed to educate New Yorkers about the risks of not controlling human activity in the New York City Watershed. In response to a Riverkeeper-planted story that ran in the *New York Post* about the closure of the Croton Reservoir because of sewage contamination, a city spokesperson claimed the closure was not due to sewage but to "organic material," prompting late-night comedian David Letterman to quip that the story "scared the organic material out of me!" A Riverkeeper "guerrilla" poster around that time depicted a water tap attached to a naked man's torso releasing yellow liquid into a drinking water glass, with the caption "Human Waste Is Discharging into 88 Percent of New York City's Water Supply."

The campaign was effective, and New York City drew up an aggressive watershed protection plan. The draft plan was leaked, and upstate communities that resided in the watershed went ballistic. They were radically opposed to any efforts by the city—against whom they held an abiding century-old grudge for commandeering control of their water—to restrict development in the Catskills, especially in light of the depressed upstate economy. Upstate developers financed a direct-mail campaign decrying the watershed protection plan and initiated a rash of lawsuits against the city. In the spring of 1995, Riverkeeper persuaded Governor George Pataki, who had just taken office, to step in to help mediate.

In 1997, after two years of difficult negotiations, a truce was reached and a Watershed Memorandum of Agreement, or MOA, was signed. The MOA had the support of the city and state, more than 70 upstate municipalities, and several major environmental groups, including Riverkeeper. The upstate communities agreed to drop the lawsuits and accept limits on growth in the watershed, while the city agreed to invest $1.5 billion in a comprehensive program to preserve land around the reservoirs and streams, construct new storm sewers and septic systems, and upgrade sewage treatment plants. The city also agreed to pay the property taxes on all city-owned property as well as provide financial and technical assistance to local farmers and foresters to help them stay on the land and manage it sustainably.

In the end, New York City ratepayers saved billions of dollars in avoided water rate hikes. As a result of this historic agreement, New Yorkers can continue to enjoy clean, tasty, unfiltered drinking water, having helped to preserve the world-renowned scenic landscapes of the Catskills, fish and wildlife habitat, wetlands, and forests that sequester carbon.

The Watershed Agreement, and the idea of protecting drinking water at its source, quickly became a model for other cities around the United States and the world. More than 140 U.S. cities have studied and adopted an approach similar to that of New York's, and in 2001 the European Union instituted a requirement that member countries protect their watersheds.

For more than four decades, Riverkeeper has used law, science, media, and old-fashioned grassroots organizing to prosecute hundreds of polluters, including major corporations like Exxon Mobil, Ciba-Geigy, ARCO, General Electric, and municipalities large and small, resulting in billions of dollars in environmental remediation and restoration.

Over that time, we have seen a dramatic improvement in water quality, evidenced by the reopening of a dozen swimming beaches, as well as the recovery of signature Hudson River fish such as the striped bass and shortnose sturgeon. Record numbers of residents and visitors now flock to the river each year to boat, bird-watch, kayak, swim, and fish.

But what excites me the most about being a riverkeeper is the larger work we do to engage the public in the fight not just for clean water but for democracy and justice. For instance, getting local residents involved in a campaign to stop a proposed shopping mall or misguided residential development helps teach them how municipal government works and how to participate effectively in local planning and decision-making. Once informed and empowered, these citizens become more civically active, redressing other injustices and forcing business and government to be more accountable.

The Hudson's rebirth is an epic comeback story and hailed internationally as a model for ecosystem restoration. Because of our success in restoring the Hudson and safeguarding the New York City Watershed, Riverkeeper has become the inspiration and model for more than 200 other "waterkeeper" programs in more than 20 countries on 6 continents around the globe, each working to protect its local waterways and drinking water supplies. This alliance of waterkeepers is arguably the fastest growing grassroots water protection movement in the world.

The environmental movement has been successful in codifying the idea that environmental protection is important to society. We also have achieved substantial gains in land preservation, air and water quality, and the control of certain toxic or dangerous substances.

But although we have scored major victories and made some significant gains, are we winning the battle to save the planet? If the goal is to preserve and restore the Earth's natural ecological balance and the services it provides so that we and other species may continue to prosper, then I would say we are losing the battle and losing it badly. Our climate is warming, our oceans' fisheries are crashing, species extinction is accelerating rapidly, rates of asthma and mercury poisoning are unprecedented, and nearly a third of the world's people don't have access to clean drinking water.

How can this be? Conservationists, starting with Sierra Club founder John Muir, have been working to protect our environment for more than a century. In my view, it is because we are a movement of incrementalists who exhaust ourselves to achieve small or temporary gains but don't have the firepower needed to bring about long-term systemic change. This is not by design, of course, but because of the nature of our political system, which favors moneyed interests over principled interests. But whatever the cause, we can no longer afford to be a movement of incrementalists when the problems we face require fundamental, even radical, changes—a veritable paradigm shift in how we live and do business.

I believe our inability to make bigger, more lasting gains stems from two major factors, among, possibly, many others. The first is that our "command and control" system of regulating harmful activity is inherently inefficient and biased in favor of polluters. A recent series in the *New York Times* details the failure of the EPA, and most of the 50 states, to adequately enforce federal laws enacted to protect water quality. The second is that we have yet to become a mass movement of citizens committed to real reform in the environmental arena. While we have had some success popularizing the notion of being "green," we have not yet inspired the deeper commitment among Americans to the policy and lifestyle changes necessary to reverse our environmental decline.

Command and control describes a common approach to environmental regulation whereby regulators determine "safe," or at least "acceptable," levels for various contaminants and then issue permits to polluters allowing them to discharge a limited amount of those contaminants into the air or water. Other command and control laws may require companies to use certain technologies that treat or limit the amount of effluent released into the environment. Most of our environmental laws are designed so that these permits are reviewed and reissued every five years, requiring ever stricter limits on emissions until pollution is eliminated altogether. In fact, the Clean Water Act, passed by Congress in 1972, had as its central goal eliminating all water pollution by 1985! Well, nearly 25 years later, more than 40 percent of our nation's rivers and lakes are still too polluted to safely fish or swim in—"fishable and swimmable" waterways being the ultimate aim of the Clean Water Act. Despite our gains in water quality, our progress is slow at best, if we're making progress at all.

So why isn't our current regulatory approach working?

First, there is a backlog of toxic substances used by industry that aren't even regulated. And for those that are, most environmental agencies are too understaffed and underbudgeted to be able to properly enforce the regulations established to control them. So groups like Riverkeeper, which exist primarily to fill the gap created by underperforming agencies, spend most of their time raising scarce resources to enforce these regulations ourselves. Even though we have a high success rate, stopping or curbing pollution in nearly all the cases we take on, we are able to pursue only a fraction of the violations being committed every day. These cases can take years before they are resolved, thus further tying up and distracting environmental groups from pursuing other polluters.

In New York State, there are more than 1,000 expired Clean Water Act permits. This means that the companies that hold them are not going through the periodic review and technology upgrading process that Congress envisioned. For most companies it's far easier and less expensive to pollute and see if you get away with it. For even if you eventually get caught, most regulators will let you off with a minimal fine, if they fine you at all, and you will have saved a lot of money putting off investing in the required pollution controls.

When we're not fighting individual polluters or raising money, environmental advocacy groups spend time lobbying agencies to strengthen or enforce certain regulations—or simply defend those same regulations against relentless attempts by powerful industry groups and their friends in government to weaken them. In my view, the current regulatory regime generally favors the status quo, which results in Americans' having to tolerate unacceptably harmful levels of pollution.

The other problem we face is that we are not a mass movement committed to fundamental change and we don't appear to be becoming one any time soon, despite the clear urgency. Still by and large a movement of well-educated, upper-middle-class people, we have failed to truly engage the vast majority of Americans in the fight for a clean environment. The irony, of course, is that we have not been able to fully engage or mobilize those who are most affected by environmental pollution—the mostly poor, mostly black and Hispanic communities where many of our most toxic industries are located.

This failure to create a mass movement for the environment has been particularly apparent in the last few years, when it looked like the issue might finally get the attention it deserves. When Al Gore's *Inconvenient Truth* came out, record numbers of people turned out to see it. It proved to be a catalyst in finally mainstreaming the issue of climate change and, it seemed, of the environment in general. And yet less than a year later, during last year's presidential election, the environment ranked 20th in a major poll as the issue Americans are most concerned about. We have overwhelming evidence that we are doing irreparable harm to the natural systems that we depend on, not only for our comfort but also for our survival, and the issue ranks only 20th?

Why is this? In my opinion, it is because when it comes to the environment, half of the country doesn't get it, or is in denial that there's a problem. Among this group are those who are too poor, misinformed, or uninformed to have the time or inclination to care about the environment, as well as those who are wealthy and well educated enough to know the importance of protecting the environment but willfully choose, for ideological or self-serving reasons, to pretend otherwise. The other half of America comprises those who understand what is at stake but are, for the most part, as yet unwilling to change their overly consumptive lifestyles. This is a very large group of us who write occasional checks to our favorite environmental group and recycle and think we're doing our part.

We shake our heads in dismay when we read about companies like Exxon or General Electric or DuPont who are doing terrible things to the environment. We throw up our hands and say, "What can we do? They're too powerful—there's nothing we can do to stop them." But we say that not because we believe it or because it's true. It's not. In fact, all the power lies with the electorate; we just choose not to exercise it. And besides, these companies are acting on behalf of their shareholders, and those shareholders are us. And they are plundering natural resources, desecrating the land, and exploiting local workers in distant places to get us the products we crave.

The reality is we don't want to stop consuming their products or, if we're shareholders, stop making money from their exploits. Our capitalist system relies on infinite growth, which means companies have to sell more and more of their products to stay in business, and thus use advertising to dupe us into thinking that we have to have their products, and those products will somehow make us happier. But we should know better. In fact we do know better, but like any addict, we can't help ourselves. We've got to have the Hummer, or the 25,000-square-foot house, or fly to Paris for the weekend.

But until "we, the people" are ready to make the transition to a more sustainable lifestyle, we should not expect the corporations and politicians who seem to hold all the cards to be our Moses and lead us to the promised land. It will happen only when we decide it has to happen. Unbeknownst to most Americans, that time has already come.

So it seems our current environmental regulatory approach is inadequate to address the challenges we face, and our countrymen and -women don't seem ready to sacrifice much to protect the planet. So how do we come up with a better way to end the pollution of our water and environment?

In my view, the most reliable and efficient solution would be to reform our perverse tax system. Currently, we tax the things we ought to be encouraging more of—corporate revenue, individual income, sales, and labor, for instance. But other than perhaps cigarettes and alcohol, we don't tax the things that are "bad" for society and we want to discourage, like pollution and fossil fuel consumption.

Taxing pollution would provide incentives for "good behavior" that are mostly absent from the marketplace. Companies that reduce their emissions would be rewarded with lower tax bills; if they reduce or eliminate their waste, they pay less or no tax at all. By internalizing the full cost of production, which includes pollution, the price of goods will reflect the actual cost of bringing them to market. If some of these additional costs are passed on to consumers, we'll respond by consuming less of the products that most pollute our environment.

While the government would still have to decide which substances would be permitted or banned, we could replace the old command and control regulatory model with a self-regulating tax structure in which eliminating pollution will be in every company's best (financial) interest. We have a workable system for collecting taxes, and already require companies to monitor and report their discharges, so it should be relatively easy to calculate emissions and levy taxes. There are details as to how you tax different types of pollution, but the basic logic of "taxing bads, not goods" is unassailable.

Skeptics will say that businesses will fight to defeat any attempt to implement a system of pollution taxes, fearing an additional tax burden. But pollution taxes could be offset with a decrease in corporate income taxes.

Using taxes to reduce pollution is not new. Other countries have been using pollution taxes for years. In the Netherlands, a tax on the emissions of heavy metals like lead and mercury resulted in a 90 percent decrease in emissions in less than two decades. A German tax on toxic waste led to a 15 percent drop in just three years. And here in the United States, the cap and trade system we implemented to curb acid rain, although an indirect tax, was spectacularly effective in significantly reducing emissions of sulfur dioxide and nitrous oxide.

Possibly the most effective use of a pollution tax would be a tax on carbon. A tax on all carbon-generating fuels would increase the price to consumers of purchasing greenhouse gas–intensive fossil fuels, thus lowering demand and emissions and forcing consumers to seek cheaper alternative sources of power. Internalizing the full costs of production—including the cost to society of climate change, dirty air, et cetera—will level the playing field and allow renewable energy producers (for instance, wind, solar, hydrogen fuel cells) to compete.

Through economies of scale, increased consumer demand for renewable power will lower the marginal costs of producing each unit of that power, further driving down relative costs. And under the leading carbon tax proposal, all of the revenue from the tax would be returned to every taxpaying American, not only providing a further incentive to consume less fuel (because of the net savings), but also ensuring that the tax doesn't unfairly penalize the poor, who spend a higher share of their income on fuel.

As I write this, Congress is considering a cap and trade climate bill, passage of which would be considered a landmark achievement. (In my view, this cap and trade approach, which places a price on carbon, is inferior to a carbon tax approach, which is more transparent and far less complicated and prone to gaming and manipulation by the fossil fuel industry.)

The cap and trade bill that has emerged from the House of Representatives, although laudable in its ambitions and certainly an improvement over the status quo, is so laden with needless complexity, loopholes and giveaways to the coal industry that its emissions targets are probably a fantasy. And yet most of the big environmental groups are out there lobbying hard to get it passed. It's not that they love the bill or don't understand its serious flaws, it's that it's all they think they can get. And without Americans pushing hard to back them up, they're probably right.

If we want change, we have to be the ones to make it. So far, despite having elected as President a self-proclaimed "change agent," Americans have not demonstrated a true commitment to change, not on the environment, health care, or any number of other pressing issues facing the country.

We have already passed the point of no return in unalterably changing how nature functions. No rainfall, no bird migration, no flap of a butterfly's wing will ever again be quite the same as it would have been had we not boosted the levels of carbon in the atmosphere. It's done. We will never again know a natural planet where the disproportionate impact of humans isn't evident just about everywhere we look. The gull is no longer just a gull, separate and apart from us humans, but a by-product, a weird amalgamation of nature and industry, a plasticized and chemically "enhanced" version of his ancestors. The six-pack ring was just the plastic we saw, but gulls and just about every species on the planet, including humans, are loaded with plastic, mercury, or some combination of other chemicals we routinely discharge into the environment. Rachel Carson's terrifying vision of a chemically altered and compromised world is the world we now inhabit.

As capable as we are of destroying things, we are capable of creating beauty and transcending our basic animal nature. I believe we still have a chance to make the transition to a carbon-constrained society quickly enough to avoid the worst consequences of global warming. And yet we have to really want it. While I remain hopeful, right now there are few signs that we do.

As a child is dependent on his mother, so are the Earth's creatures now dependent on us humans not to destroy them. We are no longer one among millions of species who co-exist and share the planet as relative equals. Humans have now assumed the role of God in deciding the fate of the Earth and all its inhabitants.

While the Earth does not "care" what happens, and will go on, with or without us, it is profoundly depressing to think that our ultimate legacy will be having wiped out millions of our fellow species and made the planet less habitable for future generations. As it is, it pains me to think that my children or grandchildren may not ever know what it's like to see a tiger or polar bear in the wild, or even take comfort in knowing that they're out there, living their lives according to nature's plan.

We've made a terrible mess, and we are duty bound to clean it up as best we can. It won't be the same planet we inherited, but at least our children and theirs will learn from our mistakes and better understand how to live in balance with nature, altered though it may be.

When I think back on that gull, I take some small comfort in the fact that it did not die entirely in vain. The image of it, tangled up in that plastic, stirred an anxiety that remains with me today, pushing me to do my small part to end our hostile relationship with nature—and thus with ourselves. When we fully comprehend that we cannot pollute the environment without polluting our own bodies and those of our children, perhaps we will be ready to make the changes we need.

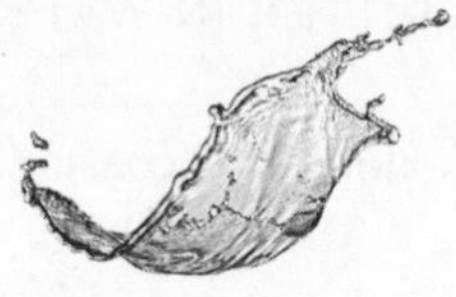

DYNAMIC FOR CHANGE

Christine Todd Whitman

Christine Todd Whitman is the president of the Whitman Strategy Group (WSG), a consulting firm that specializes in energy and environmental issues. WSG offers a comprehensive set of solutions to problems facing businesses, organizations, and governments. Governor Whitman served as administrator of the Environmental Protection Agency (EPA) from January 2001 until June 2003. She was the 50th governor of the state of New Jersey and its first female governor, serving from 1994 until 2001. As governor, she was recognized by the Natural Resources Defense Council for instituting the most comprehensive beach-monitoring system in the nation. As EPA administrator, she promoted commonsense environmental improvements such as watershed-based water protection policies. Governor Whitman chairs the Water Policy Institute, bringing together water leaders including water users, industries, water suppliers, government, and nongovernmental organizations to work collaboratively to develop innovative, sustainable solutions to address water supply and quality issues.

THE DAYS I SPENT FISHING with my father along the small trout stream on our family farm in New Jersey are among my most cherished childhood memories. I learned about the water by watching it, splashing and fishing in it, and sharing stories with my family about the big one that got away. That's where my lifelong interest in water protection began.

It has been 37 years since the adoption of the Clean Water Act. The progress America has made since that time in cleaning its water is impressive. Gone are the days when rivers in the United States spontaneously combusted and garbage was more plentiful than seashells along our coasts. Still, much remains to be done.

Today, some 40 percent of America's waters are impaired. We can do better than that. I believe the answer to further improvement lies in adopting a holistic watershed approach. Restoring the health of the watershed will be crucial in our response to climate change.

The watershed approach is not a new idea. John Wesley Powell, the explorer of the Colorado River, suggested it in 1889 in a speech to the Montana Constitutional Convention:

> I want to present to you what I believe to be ultimately the political system which you have got to adopt in this country, and which the United States will be compelled sooner or later ultimately to recognize. I think each drainage basin in the arid land must ultimately become the practical unit of organization, and it would be wise if you could immediately adopt a county system which would be convenient with drainage basins.

A watershed is an area of land that drains to a single body of water. That makes sense. But did you know that all landmasses lie within, and every person lives on, a watershed? The image of the watershed is one of the most unifying of ecological features: Picture a bowl or a funnel—depending on the type of watershed—in which all liquid that enters inevitably flows to a joint pool at the base.

There are so many reasons it is important for us to limit the pollution of our watersheds. Water is essential for survival; a clean, adequate supply is necessary to human life. As children we learn that up to 60 percent of the human body is composed of water; we hear about the importance of keeping hydrated, of washing our hands, and of bathing frequently. Outside the classroom, our early experiences show us other irreplaceable values of water beyond human consumption and hygienic needs. Many of us are lucky enough to have grown up playing in streams and ponds, as I did, or swimming and kayaking in or biking along rivers, lakes, and oceans. Nothing beats being at the beach, or, as we say in New Jersey, "going down the shore." President John F. Kennedy, who loved sailing and being on the water, said, "We are tied to the ocean. And when we go back to the sea . . . we are going back from whence we came."

There are countless watersheds in the world—2,110 in the continental United States alone—ranging in size from those that drain to small streams to those that empty out into vast oceans. Watersheds are not marked by local or state boundaries. Outside the United States, they cross international borders. Every individual within a watershed is equally dependent on the shared body of water, and every individual affects the quality and availability of that water. As I learned at a very young age, water links our communities in profound and complex ways. It is in our universal best interest to preserve the shared source of water. The watershed has the potential to become the necessary dynamic force for change, a focal point for people and communities in their desire to adapt to change with optimistic, healthful actions that preserve the ecosystem that sustains them.

But first we must understand the scope of the watershed approach, and look at the size of the "community" that is really involved. Most Americans probably think of water pollution simply as the discharge of pollutants into water. This intentional discharge is known as point source pollution, which the U.S. Environmental Protection Agency (EPA) defines as "any single identifiable source of pollution from which pollutants are discharged, such as a pipe, ditch, ship or factory smokestack." The progress we have made in improving America's water quality reflects the fact that we have largely eliminated the direct discharge of pollutants into waterways.

Most of the pollution that threatens America's waters today comes from nonpoint sources. Nonpoint source pollution occurs as water moves over or through land, thus transporting natural and man-made pollutants to bodies of water that may be many miles away. These pollutants can be the result of a major spill or something as small as the discharge that washes off the driveway after you change the oil in your car. Every eight months, as much oil makes its way to our coastal waters from nonpoint sources as was spilled from the *Exxon Valdez* in 1989. Agricultural nonpoint source pollution and the growth of paved surfaces—roads, sidewalks, parking lots, roofs—result in storm water runoff that ultimately affects streams and tributaries. All developed landmasses suffer from nonpoint source pollution, and it poses the greatest threat to America's water health today.

Sadly, because of the pollution that enters and affects every surface of every watershed, the purity of our water—along with the services it renders—is compromised. In 2004, EPA informed Congress that about 44 percent of the country's surveyed stream miles, 64 percent of surveyed lake acres, and 30 percent of surveyed bays and estuaries were too polluted for swimming, fishing, and drinking.

So we can see that unification of governmental agencies and nongovernmental organizations within watersheds is crucial to protecting our waters. But how exactly do we go about this? There are many instances of people and organizations working together to improve their water. One of my favorite examples of watershed cleanup appeared in an article by a student at the Milton Academy. He wrote about a visit with his marine biology class to the nearby Quincy, Massachusetts, beach, which received a ten-million-dollar grant from the EPA. After marveling at all of the marine life they found at the beach, he concluded his article by saying, "What's most amazing about the beach is that we students had no idea that just a few years ago . . . Wollaston used to be an environmental disaster." In attributing the remarkable change in Wollaston Beach to "the hard work of the EPA, MWRA (Massachusetts Water Resources Authority), and the Town of Quincy," the student highlighted one of the key benefits of the watershed approach: collaboration across political boundaries.

The word "watershed" was originally derived from the older term "shedding," which meant splitting or dividing. Watersheds, by this alternative definition used outside North America, are the high ridges or other landmasses that divide different drainage basins. The symbolic difference between the definitions of a watershed is significant: one is a force for unity, the other for division. The world, as seen through the hydrologically defined lens of the watershed as drainage basin, stands to benefit from more collaboration.

We need to embrace a collaborative effort for preservation, inevitably bound by the shared watershed. Through a widespread effort to reduce pollution and other strains on our waterways, we can uplift economies, strengthen neighborhood communities, and secure the health of humans, other animals, and ecosystems for years to come—and teach our grandchildren to fish in the same clear streams that we did.

THE PÁRAMOS OF AYABACA

Jorge Recharte

Jorge Recharte, director of the Andes Program at the Mountain Institute (TMI), is based in Huaraz and Lima, Peru. He holds a Ph.D. in anthropology from Cornell University. Recharte joined TMI in 1997 after spending three years in Ecuador working for the Latin American Social Science Faculty (FLACSO), designing and heading the graduate education and research program in mountain societies and sustainable development. Between 1980 and 1981 and 1990 and 1993, Recharte was an associate researcher at the International Potato Center, where he worked developing participatory research methodologies in agriculture. He currently serves on the board of the Common Good Institute in Lima, Peru, and is a member of the Andes Chapter of the International Mountain Society.

I DISCOVERED THE BEAUTY of the *páramos,* the open rangelands of the Andes, in 1995 in Ecuador, when I was taken there by my old friend Norman Simmons, retired from the Canadian Park Service. The páramo ecosystem, located roughly between 9,800 feet above sea level and the highest peaks of the Northern Andes, is a mosaic landscape dominated by grasslands and dotted with lakes, wetlands, springs, and cloud forests. At a continental scale, the páramos are an archipelago of wetland "islands" that spread, like the beads of a long rosary, along the crest of the Andes from Venezuela and through Colombia and Ecuador to the northern frontier of Peru. The fact that 60 percent of the nearly 3,000 vascular plants found in this ecosystem are endemic or unique to this biome—also the refuge of highly threatened animals like the spectacled bear *(Tremarctus ornatus)* and mountain tapir *(Tapirus pichaque)*—make this a place of unique value to the world.

Besides their value in terms of biodiversity, the páramos are priceless as natural reservoirs of water for the whole of the Andes north of Peru, all the way to Venezuela. They are perhaps the single most important natural regulator of water for hundreds of small highland towns and millions of humans in cities like Piura, Loja, Cuenca, Quito, Bogotá, and Mérida.

Since my eye-opening inaugural trip more than a decade ago, I have dedicated part of my career to this beautiful ecosystem and the people who live there. More recently, in 2004, I witnessed a grassroots commitment to conservation that was grounded in individual citizen, household, and village interest, yet at the same time was also inspired by a sense of duty to this special place of water.

My journey in 2004 began among the towns and people of Ayabaca province in the Piura region of Peru's highlands, near the border with Ecuador. Peru has historically been a mining country, but in the past 20 years mining stakes have grown exponentially to encompass nearly 32 million acres, close to 10 percent of the country's territory, thus overlapping with a large part of Peru's almost 6,000 peasant communities. Economic growth on the coast has also boomed, with double-digit annual economic gains in some regions of the country. This trend is strong, and it reached Ayabaca a few years ago. Mining stakes in the highlands of Piura were followed by large-scale exploration and, soon after, by plans to set up open-pit mines near the páramos that are the source of water for this region. The almost inevitable consequence was a conflict reflecting the clash of different worldviews and incompatible interests.

My 2004 trip to the town of Pacaipampa was at the behest of its mayor, Juan Manuel García, who invited me to visit their páramos and seek opportunities to promote their conservation. An old 4x4 truck sent by the municipality picked up my party at four in the morning in Piura, a city located in an arid desert near the Pacific Ocean. My group was invited to ride in the comfortable front and back seats; the back of the truck was packed with an impossible number of people, a typical picture in the boondocks of Peru. After 12 hours on the worst kind of roads one could possibly imagine, the rusty truck arrived in Pacaipampa. Eager to meet the mayor, I was hugely surprised when a strong, short, and merry man emerged from among the multitude of people in the back of the truck to greet me: "Hi, I am Juan Manuel, the mayor of Pacaipampa. Tomorrow we go to the páramo to visit the Huaringas." Seeing the mayor ride the truck with his fellow citizens, like one of them, making sure his visitors were comfortable, made me realize that something exceptional was happening in this town.

The Huaringas are secluded, remote lakes whose waters are said to have miraculous healing powers. They are located in the alpine grasslands near the peaks, the homeland of the last populations of spectacled bears and mountain tapirs. Traveling for almost a day uphill from Pacaipampa, we slept in a hut near the páramo before beginning the final leg of the trip early next morning; we wanted to climb to the Huaringas in the morning and be out from there before dusk.

The Huaringas region has no permanent human occupation, and is therefore silent, wild, and beautiful—a numinous site. Each time I enter these secluded valleys I have the sensation that I enter a cathedral without walls. Local residents, in fact, say it is a sin to enter in disrespect, making noise, or in large groups of people. The lakes are often visited by *curanderos* or *maestros,* shamans who bring patients to be cured by immersion in these waters. Many are local people, but in recent years and in some places (like Huancabamba south of Ayabaca) the number of visitors has increased beyond the carrying capacity of the place, creating all sorts of problems.

One of the most common ceremonies performed in the lakes is called *florecimiento,* or "flowering": a ritual that, performed in this wild setting of páramos and lakes, makes human beings whole again, lifting their spirits and charging their energy and vigor. Water, and particularly pure, virginal water, like that in the Huaringa lakes, has a central role in healing rituals. Indeed, upon our arrival in the Huaringas, our new friends from Pacaipampa performed a florecimiento ritual on the shores of Laguna del Rey. The shaman, blowing perfumed water over our heads, arms raised toward heaven, described in a whisper the splendor of the landscape: flowers, peaks, rocks, clouds, wind, each one of them inviting us all to "flower" in this setting of perfect beauty.

Shamans and patients who trek to these peaks and lakes may also drink from the San Pedro cactus. This mescaline-based psychotropic plant is said to summon a being (San Pedro) who helps people—through the loving mediation of the maestro and the spiritual force of this sacramental plant—understand their ills and see the road map to healing and prosperity.

Thus, in this first trip to Pacaipampa, I visited páramos as beautiful as, and even better preserved than, the ones I had known for a decade in Ecuador. I met for the first time practitioners of the ancestral and vigorous shamanic tradition of northern Peru, gaining the acquaintance of people profoundly motivated to protect their landscape. I also learned, through subsequent conversations, that people here linked the absence of humans, the fact that these lakes were not used, to their spiritual power to heal.

Páramos also nurture the soil. *Páramo* means "drizzle" in the local jargon. If you are standing in the valleys and people point toward the páramos, the most likely scene will be a range with clouds sitting on the peaks. "*Mira, está parameando*"—"See, it is drizzling." Levels of precipitation approach five feet per year, and relative humidity ranges between 65 percent when "dry" and the saturation point; it is also very cold because of the high altitude. This means that plant evapotranspiration (loss of water from the soil both by evaporation from the soil surface and by transpiration from plant leaves) is low. All these conditions contribute to make "páramo" and "wet environment" synonyms. Páramo soils—volcanic in origin with high levels of carbon—have extraordinary hydrological features: They can absorb water easily, retain it during the wet season, and then release it very slowly as weather "dries" (although it rarely dries totally). The end result is that páramo ecosystems have an extraordinarily high rate of hydrological yield: a relationship of flow to precipitation of 0.65. This magic number means that páramos yield more water than tropical cloud forests, though some of this happens underground, invisible to our eyes.

The Peruvian government issued land grants in the early 1970s as part of a national agrarian reform. Since then, with steady population growth and the need to feed themselves, farmers have increasingly deforested the hills and slowly moved the agricultural frontier up toward the páramos. Should they reach this zone and plow it for farming, the unique hydrologic features of these fragile soils would be affected and probably lost.

In almost three decades of moving around these mountains I have witnessed the erosion of traditional knowledge, crop diversity, and lands and the fading away of a rich tradition of dances, tales, and other expressions of culture that embody sense of place. In Peru, my home country, *progreso* has been a banner raised high by rural communities themselves since the dawn of the 20th century. Community leaders and peoples have painfully carved out with their own sweat, money, and blood their right to set up schools, roads, and electricity and to exercise their rights to full citizenship. I spend my life writing projects—the arid language of rational interventions—for mountain communities who seek to preserve the best of their country and improve their livelihoods. Yes, there are wonderful local stories of determination: community-based tourism along the Inca Road, protection and restoration for ancient cloud forests or grasslands, or households that improve their income, thus allowing them to continue their mountain way of life. But unfortunately, these are small islands of possibility in an ocean of great global transformations characterized by short-term horizons, rapid growth, and erosion of all that makes each town and place in the mountains unique.

So, here I was after this trip, at the invitation of local communities, once again ready to write in the dry language of technocracy a project proposal explaining the unique value of an imperiled ecosystem. The villages that belonged to Juan Manuel's district were eager to find ways to protect their páramos, both for their natural beauty and as sources of clean water. They wanted to help people identify new means of livelihood or improve what they already had so that they stopped moving upward to the páramos. Chief among their worries was the fact that mining companies holding concessions near their páramos were exploring for minerals. From my perspective and interest in páramo conservation, I could understand their concerns: The hidden treasure of páramos was water, and without clear guidelines and understanding among government agencies and the public, this natural treasure was at risk.

The dry but necessary process of project language and structures was the beginning of an organized effort to protect this treasure. In the years since this story begins I have witnessed the dedication of the mountain villages in Ayabaca to the conservation of páramos. Communities established targets to discuss their vision of the land, identified their family and hamlet objectives for the future, mapped opportunities to manage their lands in better ways, and translated their knowledge of their territory into geographic information systems. They worked hard to strengthen their skills at local planning and to improve their farming businesses in the valleys, and they also became better at presenting their dreams and ideas in the public arena. Years later, villages have completed their vision of healthy páramos preserved as wilderness places and providers of clean water, and are now working hard to make their dreams a reality.

Although I admire the dedication of people living in remote and materially poor places to their dream of preserving water and páramos, the more striking lesson for me was to see how in village upon village in the Ayabaca highlands a movement developed to end the advancement of mining concessions in their territory through their own grassroots organization, the *rondas campesinas.*

We may glimpse in Ayabaca the predicament in which so many corners of the globe that are rich in natural resources find themselves. Nations need to grow economically, but often, unfortunately, they do so at any cost and as fast as possible. Remote rural communities like Ayacaba offer an alternative perspective on the future of natural resources. Local opposition to mining is grounded on an intuitive wisdom of conservation. It also implies a long-term vision in spite of acute material needs. Ayabaca is a story of hope because the clash of these two visions—which at times has been violent—is slowly taking a direction that I think could become an example of care for the earth and water, should all the necessary pieces continue to fall into place.

I learned several lessons that continue to inspire my enthusiasm for conservation of páramos. First, peoples who have historical and cultural roots inscribed in their landscapes, like the people of Ayabaca do in their páramos, have a profound wisdom about how to care for the land. This is often intuitive knowledge, not necessarily all of it neatly grounded in science and objective facts. Yet it should be taken very seriously as the standards by which we—people who do not belong to the place—should also see the future of the land, whether we are conservationists, development workers, miners, builders of dams, or government representatives. Trying to understand the world through the eyes of those who live in the place has supreme importance.

Second, I have learned that when local communities are able to articulate this knowledge (and the values that underlie it) in the language of urban people, then doors to communication open up. If their values are respected and listened to, and if their perspective on the future of the landscape and water is at least understood, then these small communities begin to develop a sense of inclusion and common ground with their urban counterparts. This has started to happen in Piura, where highland folk traveled, for the first time, from their hamlets to the coast, where they presented their plans to manage the páramos water sources to politicians and the public. The Piurans were also able to explain why it was in the best interest of those living on the dry coast to listen to their pleas. The gap between these alternative worldviews is probably irreducible, but a willingness to communicate should always be welcomed.

Finally, water is a resource of unique value for highlighting common interests among citizens. It creates opportunities to build democratic habits. Water opens doors to long-term and more inclusive visions of what it means to be part of the same nation or even the same planet—the only one we have.

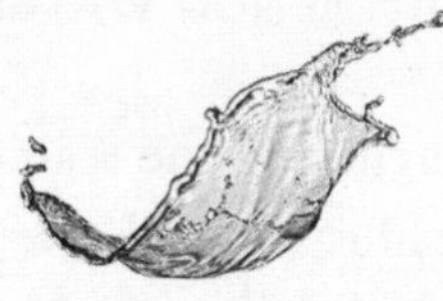

TENDING THE LAND

Frederick Kirschenmann

Frederick Kirschenmann, a longtime leader in national and international sustainable agriculture, shares appointments as distinguished fellow at the Leopold Center for Sustainable Agriculture at Iowa State University (ISU) and president of Stone Barns Center for Food and Agriculture in Pocantico Hills, New York. He also oversees management of his family's 3,500-acre certified organic farm in south central North Dakota and is a professor in the ISU Department of Religion and Philosophy. Kirschenmann holds a doctorate in philosophy from the University of Chicago and has written extensively about ethics and agriculture. He has held numerous appointments, including the U.S. Department of Agriculture's National Organic Standards Board and the National Commission on Industrial Farm Animal Production operated by the Johns Hopkins School of Public Health and funded by Pew Charitable Trusts. He is currently dividing his time between Iowa and New York, to explore ways that rural and urban communities can work together to develop a more resilient, sustainable agriculture and food system.

I LEARNED THE IMPORTANT LESSONS about water very early in my life. My father and mother began their life on our family farm in North Dakota after they got married in 1930. Their early years as beginning farmers were thus spent in the midst of the Dust Bowl, an experience that made an indelible impression on my father. Although the Dust Bowl was clearly about water (in this case, the lack of it), my father understood intuitively that the devastation that it wrought on our land was not solely about the lack of water, it also was about the way land was farmed. The weather, including the scarcity of rainfall, was the immediate cause of the Dust Bowl; however, the farming methods of that era had left the land vulnerable to incredible soil loss. As a result my father became a radical conservationist, and from the time I was five years old I can remember him lecturing me, pointing his finger at me, admonishing me to "take care of the land." As far as he was concerned, taking care of the land was the most important moral duty imposed on any farmer—not only for the sake of the land, but also for the economic survival of the farmer.

Consequently, water has never been an isolated "thing" for me. I understood from my father's tutelage that water was only one part of a complex web of living relationships that included, among other things, soil, climate, biodiversity, and husbandry.

Growing up on our North Dakota farm helped me to further appreciate the complexity of water's role and the importance of its relationship to everything else on the farm. In central North Dakota, average annual moisture amounts to 14 inches, which made me very aware that life depended on precious, timely rainfall, especially since there were no sources of water for irrigation in this part of the state.

I also noticed, early on, that perennial grasses in our native prairie fields were much more resilient under drought conditions than our annual crops. That raised several questions for me. Could we design alternative cropping systems that would perform better under uncertain water conditions? Were there ways to manage our soil so it would absorb and retain more moisture to help sustain our crops during drought periods? Could we design a farming system with sufficient diversity to increase our chances of surviving times of drought?

I'll return to these questions later.

By the time I went off to college in the 1950s, industrialization had reshaped U.S. farm life. The industrial model radically transformed all of agriculture in just a few decades.

Like other industrial economies, this new version of agriculture concentrated on a single goal—maximum, efficient production that fueled short-term economic returns. Ecological economist Herman Daly has reminded us that such industrial economies function as if they were not bounded by the limits of the ecosystems in which they exist. Industrial agriculture is a prime example of such unfettered behavior. It assumes that the resources that sustain industrial agriculture (fossil water, fossil fuels, soil) are infinite and that the sinks in nature that absorb its wastes are equally limitless. Today we are rapidly becoming aware of the limits of both.

It was not until I went to graduate school that I began to understand the intellectual culture that informed this industrial mindset. There I learned that the industrial economy was based on a philosophy of science that reached back to the 17th century. From this scientific perspective, nature was a collection of mechanistic objects that humans could manipulate purely for their own benefit.

This view of the world profoundly affected the way we lived and (once adopted by agriculture) the way we farmed. As in other industrial economies, the singular goal of efficient, maximum production led us to develop an agriculture grounded in specialization, simplification, and economies of scale. It led us to an input/output system of agriculture that required enormous inputs (fertilizers, pesticides, farm equipment, et cetera) and paid scant attention to the limits of those resources or to the impact they might have on the environment. Based on this 17th-century scientific construct, we assumed that we could isolate discrete parts of nature and modify them with minimal effects on the functioning whole, behaving as if there were no interdependent organisms, no interacting networks, no unintended consequences. Of course, all of these assumptions were contrary to the emerging sciences of ecology and evolutionary biology and the science of networks.

Although the science of ecology has been evolving for decades, it has barely begun to influence the agriculture paradigm of the 21st century. We still manage farms as if all of the parts of the farm, including water, are separate entities. However, that method of farming is becoming increasingly dysfunctional, and the philosophy that informs it is being questioned more rigorously.

Indeed, cultural historian Morris Berman asserts that there is something "wrong with our entire world view." Berman points out that since the dawn of the scientific revolution we have gradually adopted a "mechanical philosophy" that "insists on a rigid distinction between observer and observed" and assumes that our personal well-being is contingent upon acquiring personal wealth through the exploitation of natural resources. In this worldview, "the acid test of existence is quantifiability, and there are no more basic realities in any object than the parts into which it can be broken down . . . atomism, quantifiability, and the deliberate act of viewing nature as an abstraction from which one can distance oneself." These attitudes all directed us toward the notion that English philosopher Francis Bacon suggested was "the true goal of science: *control.*"

Furthermore, our attempt to isolate the welfare of the human species from the health of the rest of the biotic community is a direct outgrowth of this worldview. And perceiving water as if it were a separate entity, a thing, a commodity, is part and parcel of this same compartmentalized scientific culture.

The notion that all of our problems have a technological solution is an additional extension of this deeply entrenched culture. Furthermore, given our technological prowess, we have now convinced ourselves that we are in charge, that we are in control, that we can reinvent nature, that we are the end point of evolution. Industrial agriculture is simply an extension of that worldview.

But we now know that nature is *not* a collection of objects. It is *not* a machine. We are *not* the end point of evolution. And we are *not,* as environmentalist Aldo Leopold reminded us, "conquerors" of the land community, we are simply "plain members and citizens of it." Furthermore, we are gradually learning that farming as a command and control enterprise in this interconnected world is subject to enormous unintended consequences.

In our new view of science we see that the foundation of all life, as biologist Lynn Margulis and writer Dorion Sagan have reminded us, is the microcosm—the "trillions of communicating, evolving microbes." We are the latecomers to the evolutionary train. "The visible world is a late-arriving, overgrown portion of the microcosm, and it functions only because of its well-developed connection with the microcosm's activities." Nature figured out how to sustain itself in the face of environmental aggravations long before we arrived, and the survival mechanisms that nature evolved were deeply entrenched in its microbial memory. Furthermore, nature has gone through a costly process of natural selection to arrive at the present moment. "More than 99.99 percent of the species that have ever existed have become extinct," said Margulis and Sagan. In the light of this evolutionary history, they wrote, "it is difficult to retain the delusion that without our help nature is helpless. As important as all our activities seem to us, our own role in evolution is transient and expendable in the context of the rich layer of interliving beings forming the planet's surface." And these evolving microbes can easily undo the most inventive and technologically sophisticated command and control strategies of any farmer.

In order to truly understand how water is written into our world, our lives and our agriculture, it is therefore necessary to view it in the context of this living ecology. Fresh water is one of the natural resources that we have been exploiting at a perilously unsustainable rate in our industrial food and agriculture system. Renowned environmentalist Lester Brown reminds us that on average we each drink about four liters of water per day, "while the water required to produce our daily food totals at least 2,000 liters." Seventy percent of that water is used for irrigation, twenty percent is used by industry, and ten percent for residential purposes, according to Brown. And it adds up. Journalist Paul Roberts reported that, by some estimates, there are 5,169 gallons of water embedded in every pound of grain-fed beef. This is simply unsustainable by any measure.

The fresh water to produce our food comes from two sources, surface water and aquifers. Both are being drained more quickly than nature can recharge them. Brown cites several examples. Africa's Lake Chad, which is surrounded by some of the world's fastest growing populations, "has shrunk by 95 percent since the 1960s. The soaring demand for irrigation water in that area is draining dry the rivers and streams the lake depends on for its existence." Aquifers in China and India are being drained much faster than they can be recharged, leading to substantial declines in production. In parts of Texas, Oklahoma, and Kansas (which get their water from the massive Ogallala Aquifer), water tables already have dropped by 100 feet, depriving some farmers of water for irrigation. In sum, the mighty Ogallala has been drained by half since 1960!

The effects of climate change likely will add further stress to our global water resources. Most climatologists expect that one of the consequences of climate change will be more unstable climate conditions—more droughts and more floods. It is also likely that snowpack, which provides the irrigation water for many agricultural regions, will diminish as a result of climate change. For some regions, the disappearance of the glaciers could be devastating. *National Geographic* journalist Joel Bourne wrote that "Himalayan glaciers that now provide water for hundreds of millions of people, livestock, and farmland in China and India are melting faster and could vanish completely by 2035. In the worst-case scenario, yields for some grains could decline by 10 to 15 percent in South Asia by 2030."

There is ample evidence why water cannot reasonably be considered a "thing," an isolated entity, unconnected from us or from the rest of the living ecology of nature and our participation in it. The water issues we are facing, both depletion and the deterioration in quality, are tightly coupled to a complex, interconnected set of relationships, making it unlikely that we can solve our water problems apart from addressing comprehensive ecological health.

In the first place, one of the reasons that we are using such large quantities of water for irrigation is that we have not paid attention to the biological health of our soils. Soil also is not a thing, but a dynamic web of relationships with billions of microorganisms at the base of soil life. Throughout most of the industrial era of agriculture, we have treated soil as if it were nothing more than a material to hold plants in place while we inserted the synthetic nutrients plants required. We ignored the biological life of the soil. Restoring the biological health of the soil can dramatically increase the soil's porosity and therefore its water absorption and retention capacity. These soil qualities can dramatically reduce the need for irrigation and sustain plant growth during drought periods.

It was the availability of cheap fossil energy that enabled us to simplify our production systems, merely inserting synthetic inputs into the soil rather than attending to the return of organic matter by recycling all wastes to the soil—one of the most effective ways to restore soil health, according to scientific findings published in *Nature.* As the era of cheap energy ends, it may well become imperative for us to relearn how to properly manage soil.

Rattan Lal, soil scientist at Ohio State University, has, in fact, suggested that the restoration of soil health is the best option we have for addressing the combined challenges of food security, climate change, water depletion and energy demands. His ten principles of sustainable soil management explain that although proper soil management is a biophysical process, it is tightly linked to social, economic, and political determinants. He points out that "when people are suffering from poverty, they pass that suffering on to the land. The stewardship concept is important only when the basic needs are adequately met." Lal makes it abundantly clear that soil is not only part of an ecological web of relationships, but also part of a social web.

Lal further notes that "you cannot take more out of the soil than what you put in it without degrading its quality." Lal is warning us of the impending disaster of ignoring yet another web of relationships where soil is concerned. In a recent issue of *Crop, Soils, Agronomy News,* Lal asserted that using crop residues as a source of renewable energy is "essential to preserving soil quality. In addition to controlling erosion and conserving soil water in the root zone, retaining crop residues on the soil is also necessary for recycling nutrients [and] improving activity and species diversity of soil micro- and macro-fauna." In such soil, fertilizers have greater efficiency, there is a more stable crop yield, and the soil's organic matter content remains more constant.

Inevitably, managing soil in accordance with this "law of return" will require a more diverse, complex agriculture. The more we specialize and simplify agriculture, the more difficult it is to return the proper mix of wastes needed to feed the life in the soil.

The philosophy behind these insights, which I acquired in graduate school, helped me to understand the roots of the culture behind our industrial agriculture and to comprehend its weaknesses. It became increasingly clear to me that the essential resources on which industrial agriculture depended, especially cheap energy, surplus water, and stable climates, were rapidly disappearing, which suggested that its future was, at best, uncertain. The need to reexamine our industrial culture and to explore alternatives for our future food and agriculture system became a matter of urgent concern for me.

In 1976, after my father had a mild heart attack, I decided to leave academic life and return to manage our family farm operation. This provided me with the opportunity to explore alternatives to industrial agriculture in the real world of farming.

Being on the farm with full management responsibilities for the first time gave me the opportunity to address the theoretical questions I entertained earlier but now saw in the context of a real working farm: Were there ways to manage soil so it would absorb and retain more moisture to sustain crops during drought periods? Could I design a farming system with sufficient diversity to increase its resilience? Was it possible to devise a farming system that was less energy intensive? Was it possible to create a farming system that was more self-renewing and self-regulating?

In addition to his passion for taking care of the land, my father was a progressive farmer, and he had always been interested in exploring technical innovations. His goal was to be the best wheat farmer in Stutsman County! Consequently, when synthetic fertilizers first became available in our community in the early 1940s, my father was intrigued. He was deeply interested in increasing his wheat yields, and this seemed like an efficient way to do so.

But he also was concerned about the effect such inputs might have on his land and he decided to check with our county extension agent and with other farmers whose judgment he respected. Everyone with whom he conferred assured him that synthetic fertilizers would *not* have a negative impact on the health of his land. They all told him that the inverse would be true—enriching the nutrient density of the soil with synthetic inputs could only *improve* the health of his land. Based on those compelling arguments, my father became the first farmer in our township to begin applying synthetic fertilizers. The results, of course, were spectacular.

From the day that wheat seedlings emerged from the soil, until the day of harvest, one could tell exactly where fertilizer had been applied—the wheat was more robust, the yields were significantly higher. My father was ecstatic.

Of course, with this new technology he could plant wheat in successive years or grow it in simple rotations. And since wheat was the best possible cash crop in our part of the world, it simply made practical sense to raise more wheat and abandon other crops.

Replacing complex rotations with monocultures increased weed pressure. The more often we planted a cool-season crop like wheat, the more often cool-season weeds would produce seeds. Weed pressure built up over time. So my father had to begin applying herbicides to assist with weed control. By the time I returned to manage the farm, it was a fairly specialized wheat and sunflower monoculture farm operated in accordance with typical industrial farming practices—and the quality of our soil was significantly impaired.

We rarely saw an earthworm. Organic matter had declined, and the physical character of the soil had deteriorated. Soil granules had broken down, and there was little pore space in the soil. All of this meant that the soils on our farm were absorbing and retaining much less moisture from our limited rainfall. We were more vulnerable to droughts.

I remembered that almost ten years earlier, during my academic career, I had the good fortune to meet a student who had served as a research assistant to an extension specialist at the University of Nebraska. The extension specialist had designed a research project to determine the effects organic management had on soil quality when compared with those of conventionally managed fields. The student shared some of the results: The soil in the organic fields became more porous, its organic matter increased, and earthworms were present in much greater abundance. Inspired by those results and the information I gleaned from works like Sir Albert Howard's *The Soil and Health,* F. H. King's *Farmers of Forty Centuries,* and other soil science classics of the first half of the 20th century, I decided to convert our farm to an organic operation.

Making such a transition in the 1970s was challenging. There were no mentors to call on for advice, and there were very few farmers in our part of the world with any experience in making such a transition. I learned how to make compost from Bob Steffen, a farmer in Nebraska. David Vetter, my former student, helped me think through crop rotation strategies. I made plenty of mistakes. But eventually, I devised a crop rotation that helped us control weeds, recycle nutrients, reduce disease, and find a niche for our crops in an emerging organic market. We began to see the quality of our soils improve.

By 1988, when we experienced one of the severest droughts on record in North Dakota, our soils had begun to absorb and retain enough moisture that it sufficiently sustained our crops. Under those severe drought conditions, our fields still managed to produce a 17-bushel-an-acre average yield while conventional fields around us dried up, yielding no harvest at all.

Despite those results, our farm, in my judgment, is still far from "sustainable," given the challenges we are likely to see in the decades ahead.

In addition to managing my farm in North Dakota, it has been my privilege during the past decade also to be part of a community of scholars and practitioners at the Leopold Center for Sustainable Agriculture at Iowa State University, and more recently the Stone Barns Center for Food and Agriculture in Pocantico Hills, New York. In both centers we are addressing some of the rural and urban food and agriculture challenges that we are likely to face in the decades ahead. I continue to try to apply what I am learning on our own farm in North Dakota.

As I see it, the key challenges we will be facing in the decades ahead are to continue producing an adequate amount of healthful, nutritious food for a growing population in the face of disappearing fossil fuels and fossil water, declining biodiversity and genetic diversity, and more unstable climates. In an effort to anticipate these challenges on our own farm in North Dakota I have tried to frame the daunting task before us into a self-evident question: Let's assume that ten years from now crude oil will be $300 a barrel; that our planet will have only half the amount of fresh water available for food and agriculture enterprises; and that we will have twice as many severe weather events, droughts, and floods. What kind of agriculture can we put on the landscape on our farm to remain productive under those circumstances?

It is clear to me that the methods currently employed on our farm, notwithstanding some of the benefits of the organic management practices we've instituted, still will not prepare us to meet such a daunting challenge. The farm has to be redesigned to be much more resilient under such difficult impending circumstances. What do we need to do now?

In the short run, we plan to increase the presence of perennial grasses and legumes in our crop rotations. Perennial plants are much more resilient than annuals, and are likely to perform better in our new, less hospitable world. Consequently, we are in the process of converting part of our annual crop acres to perennial grass mixtures. Perennial plants have much denser and deeper root systems, and therefore obtain more nutrients and moisture from much deeper in the soil profile. Perennial plants also do a superior job of restoring and maintaining the biological health of the soil so that it absorbs and retains more rainfall in the soil profile. That allows more rainfall to be transferred back into the aquifer rather than running off the fields into streams and ultimately contributing to coastal dead zones. This means that we will slightly shift the balance of our farm's production to raise more livestock and less grain, but we will continue our current practice of *not* feeding *any* grain to our livestock. All of our grain production will continue to be sold directly into organic markets for human consumption. Our livestock will graze on the perennial grasses during the summer and feed on the forages harvested from our legumes during the winter.

In the long run, we will continue to stay tuned to the work that has been initiated by The Land Institute in Salina, Kansas, and more recently adopted by several universities. The Land Institute has been developing perennial varieties of grain for several decades. As these perennial seeds become more commercially available, we hope to convert the annual monocultures on our farm to perennial polycultures.

Perennial crops are more resilient to climate change; sequester more carbon, since they are "alive" for a much longer part of the growing season than annuals; reduce water runoff by virtue of their living cover and improved soil porosity; and dramatically reduce nitrogen leaching. All of the ecosystem services inherent in perennials make them good candidates to provide many functions our farm will need to maintain its long-term productivity.

In our local ecosystem, animals also will continue to play a significant role in our farm operation. One-third of our 3,500-acre farm is still in native prairie and will remain so, to be used for grazing by our animals in the summer months.

We will, of course, rely on the "no waste" policy that we have adhered to for the past 30 years. It was Sir Albert Howard who first alerted me to the principle of the "law of return" which requires that there be "no waste." We will continue to screen the foreign material, weed seeds, broken and shrunken kernels, et cetera, that are not suitable for human consumption from our grain at harvest time and feed this "waste" to our livestock as a source of extra protein during the winter months. The manure and deep litter (waste) that accumulates in our livestock's winter loafing areas will continue to be composted and returned to the land to improve the biological health of our soils.

Additionally, we plan to search out more innovative production systems that are based on energy exchange instead of energy inputs. We are trying to learn from creative farmers like Joel Salatin who have developed complex, synergistic systems in which the waste of one species becomes the food (energy) of another. We know from the laws of thermodynamics that simplistic systems tend to be entropic (a lot of energy flows through the system), while complex systems tend to keep more energy nested in the system by virtue of the energy exchange that takes place among the diversity of species in the system. These are the models we intend to emulate in our post–fossil fuel future.

As much as possible, I plan to be involved in my work beyond my own farm and continue to be part of the larger effort to transform our food and agriculture system. To that end I hope to champion more advances in urban agriculture initiatives. Urban agriculture has been evolving rapidly in recent years, and many creative farmers are developing incredibly productive, synergistic systems. Will Allen's Growing Power farm in Milwaukee is a prime example. By creating multiple synergies among species, Allen manages to "provide healthful food to 10,000 urbanites" on three acres of land. For example, Allen has created huge fish tanks in the center of his greenhouses that are three feet wide and four feet deep, extend the full length of the greenhouse, and are stocked with tens of thousands of perch and tilapia. Above the fish tanks Allen has installed beds of watercress. The water from the fish tanks is pumped into the watercress beds. The watercress cleanses the water for the fish, while the nutrients for its own growth are supplied by the droppings of the fish. The fish sell for $6 each and the watercress for $16 per pound.

Equally promising models of synergistic production are being developed by individual farmers in many parts of the world. These models are well suited to community food systems where small-scale farmers have found ways to produce incredible amounts of food on limited acreage for local populations. The efficient recycling of water often plays an integral part on these farms.

As our energy-water-climate challenges impose themselves upon us, we will no longer be able to divide our world into two isolated domains—rural areas that are expected to supply all of the raw materials for our food on large-scale, monoculture tracts to be transported long distances to urban centers, and urban areas reserved for centers of development, manufacturing, and consumption. We will need to gradually embrace the concept of "foodsheds"—a concept borrowed from our knowledge of watersheds. Foodsheds are geographic areas wherein people engage in a civic exercise that determines the most sustainable food system for their region. The first priority of a foodshed is to produce as much of the food as possible *by* people in the foodshed *for* people in the foodshed; exports and imports become the second priority. The concept of the foodshed has been around since at least 1929 when Walter Hedden published *How Great Cities Are Fed*. But the idea did not capture much public attention until the food crisis of the past two years.

Foodsheds and urban agriculture are highly compatible concepts. Foodshed language has begun appearing in conference literature this summer, and Scott Stringer, Manhattan borough president, has recently become a strong advocate of the concept as a way of addressing food challenges in New York City. Stringer has set a 100-to-200-mile distance as a *preliminary* context for exploring the development for a foodshed for New York City. As a way of getting the concept off the ground, Stringer is proposing that "within this area, designated growers of healthy food would be provided special access to New York City food markets and other retail and wholesale outlets, including farmers' markets. They would also be entitled to compete for a government set-aside through which schools, hospitals and other municipal institutions would be mandated to purchase a certain percent of their produce." For example, according to Stringer, "twenty percent of the annual budget for the city's $435 million school food programs alone would mean an investment of more than $80 million in producers within the New York City foodshed."

Interestingly, Stringer's office tells me that he is getting more positive feedback on his position on food, in the form of emails and letters from his constituents, than any other issue he is championing. All of that indicates that a food revolution of sorts may be under way in our country, and it may well be launched from our urban communities.

This community foodshed concept also is fully compatible with the United Nations' new mandate to foster "food democracy, food justice, and food sovereignty" as the means by which global food problems can best be solved. It also is in accord with the G8 countries' recent shift of focus for solving the problem of our growing global food crisis. The G8 now recognizes that it is a critical task to revitalize the food production capacity of local communities rather than encouraging the producing and shipping of food to such communities from other parts of the world.

How the next chapters in the story of water are written in this country and around the world will depend in large measure on how creative water use is embedded in the ecology of these new food systems.

A WALK ALONG THE RIVER

Dave Rosgen

Dave Rosgen, P.H., Ph.D., is a registered professional hydrologist with Wildland Hydrology. With 44 years of experience in stream morphology, restoration, river assessment and modeling, fish habitat enhancement, and other river studies, he designs, supervises, and monitors large-scale river restoration projects throughout the world. Rosgen also conducts short courses nationally in river morphology, assessment, monitoring, and restoration using natural channel design methods. He was the recipient in 1993 of the U.S. Environmental Protection Agency's Outstanding Achievement Award for research and development of technology to restore impaired streams. His contributions and innovations in water resources have been written about in *Time, Science, National Geographic,* and *5280.* Rosgen has written more than 50 papers in research journals, symposia proceedings, and federal agency manuals and is the author of the books *Applied River Morphology* and *Watershed Assessment of River Stability and Sediment Supply.*

RIVERS HAVE A MAGICAL attraction for those of us who are curious about how they work and who appreciate their value. As a young boy, I had my first encounters with rivers fishing the undisturbed, backcountry streams in northern Idaho in the late 1940s and early 1950s. The Westslope cutthroat trout fishery produced a fine catch along tributaries of the North Fork of the Clearwater River like Isabella, Skull, and Quartz Creeks, where the only access was by miles of horse trails. I grew to love being on the river and fly-fishing in its waters.

In the years between high school and college, I returned to these familiar rivers after being assigned to the Clearwater National

Forest in 1965 as a watershed forester. My first day off work was spent heading to "the ole fishing hole," revisiting the streams I had fished throughout my younger years. But instead of packing in horseback, I could drive up the North Fork. What I saw there first made me think I was lost and that these disturbed streams had to be different from the ones I had fished as a kid. However, upon checking the map, I discovered that these were indeed the same streams—but they had changed. The pools that used to produce many large cutthroat trout were now filled with sand. The river rocks were not slick as they used to be, and I could easily cross the streams that had previously turned me upside down. The most disturbing reality was that I got skunked—I caught not one fish!

Not only had roads been built up the North Fork of the Clearwater River and the major tributaries, but 800- to 1,200-acre clear-cuts (land stripped of all trees) were located back-to-back. Miles of logging roads spaced 300 to 400 feet apart had been constructed in the clear-cuts and across the slopes of the watershed. The horse trails I had ridden 15 years earlier were now prone to natural slumping as a result of the unstable soils along the rivers I had fished. The roads caused frequent, large landslides, and erosion was blatantly evident: Eroded soils on these steep slopes fell directly into the streams below, filling them with sand.

The clear-cuts also increased flooding, because there were no longer trees to use much of the water, which made more water available for runoff. The erosion from the increase in water in the streams and the impacts of the roads produced excess sediment and sand that accumulated downstream, damaging the fish habitat.

My startling observations that day led to a deeply concerned and troubled yet rapid return to the ranger station. The same streams that I had fished and loved as a young boy were changed, I was certain. Having seen the drastic difference, and hoping to do something about it, I approached the district ranger to communicate what I had seen and describe how this unsustainable method of timber harvest was damaging the water and its aquatic life. The forest officer countered by stating that any changes I observed were due to Mother Nature and were not the result of forest operations. I knew he was wrong. He didn't have the advantage of my vivid recollection of the character of the streams and how well they had fished, even on years following floods. However, without data, I was unable to convince him otherwise.

In these early years of my career, I was often frustrated by my inability to persuade those in charge to modify the land use practices that damaged watersheds and streams. Unfortunately, there were no funds or technology available then to repair or restore the growing number of affected streams. This led me to a career-long dedication to recording, documenting, and monitoring rivers and to studying the causes and consequence of changes in river behavior. I was able to collaborate with other hydrologists in the Forest Service who were experiencing similar frustrations, and together we developed models that would allow river managers to predict watershed issues far sooner than had previously been possible—maybe even in time to avert long-term damage. Yet this still wasn't enough. We needed to be able to show local governments and planners—through hard data—how certain activities would affect rivers so that they could change their harmful practices.

Although well intended, many traditional federal, state, and local river practices at this time severely damaged rivers. For example, in the interests of flood control, engineers widened and added levees to rivers, building continuous berms along the channel banks to prevent floodwaters from overflowing into the floodplain. However, the overwidening meant that excess sediment filled in the riverbed, actually *increasing* the flood hazard.

Sometimes water was channeled into "conveyance canals" that were hardened and lined with concrete, gabion baskets (rock-filled cages), interlocking blocks, and similar materials. Redirecting the water so that it ran counter to the winding "rules of the river" led to great loss at great cost; even with the channel hardening, managers needed to dredge the river repeatedly and repair levees. Although these traditional river works may have met some short-term, single-purpose goals, they were directly responsible for extensive damage to stream channels, water quality, and ecological function.

Unfortunately, the priorities in those years were directed toward planning the next project rather than evaluating the performance and system response from the last one. Without ongoing monitoring and an understanding of consequence, it was unlikely we'd avert disaster before it was too late. Blazing new trails is not valued when you already have an expedient and well-traveled path. The real challenge is to understand where that path is leading rather than trying to redirect the traffic. I learned one thing—change cannot be accomplished without controversy.

My first career with the Forest Service as a watershed forester and hydrologist spanned 20 years in Idaho, Montana, and Colorado. As a field man, I tried to help prevent river and watershed impairment and to develop better ways for predicting problems. One of the most urgent projects I confronted started in 1968, as the West Fork of the Madison River dumped disproportionate amounts of sediment into the main stem Madison River during snowmelt runoff periods. This turned the Madison River brown and brought the fly fishermen to the Forest Service office demanding an explanation for their rapidly declining take. They passed the buck to me to find out where the sediment was coming from and whether it was stemmed from geology or land uses. I gladly accepted the challenge but soon realized my eagerness outweighed my experience. The work required covering more than 100 miles of remote, inaccessible rivers and tributaries, much of it on horseback.

When the snow melts, the water rushes swift and deep through some sections of the river. Since there was only one bridge across the river at the mouth, any crossings during snowmelt runoff were by horseback. Eager to get to the upper watershed to measure sediment, I tied my sample bottles and sampler to my saddle and headed up-country. At a particularly unforeseen deep, fast water crossing, I felt my horse slipping downstream with the current. I turned the horse downstream while I was still in the saddle but felt him being swept off his feet. As he struggled frantically to keep his head above water, I bailed out of the saddle, holding the saddle horn with one hand and trying to swim with the other. After fighting the fear and the ice-cold raging river for more than a quarter of a mile, we finally grabbed enough footing to clamber out of the river. The next order of business was to make the cold, wet ride up to the upper West Fork cow camp where I could take a dip in the natural hot springs, warm up, and dry out my clothes. After this memorable adventure, and after searching my mind for a better method, out of necessity, I came up with a new idea.

The next year, we abandoned the horse-and-buggy era for the jet age, flying the entire 100 miles of main stem and tributaries by helicopter and using a handheld aerial camera with color infrared film. By flying and landing at various locations during the peak snowmelt runoff season, when the West Fork was dumping its muddiest load into the Madison River, we were able to get sediment samples over a wide range of concentrations throughout the watershed. We saw that the accelerated erosion was caused by overgrazing, which had depleted the willow vegetation and made the streambanks more susceptible to erosion when the melting snow caused the river to run high.

As a result of my findings, the ranger district instituted a deferred rotational grazing strategy to help the willows recover. In addition, a planned timber clear-cutting operation in the upper basin was changed to a selective harvest—in which loggers removed only half of the stand of mature trees—to combat streambank erosion. These data and our subsequent conclusions began to spur other changes in traditional land use practices. This was the turning point in my career.

As in any science, some of the answers I was learning led to more questions. I knew that I needed to seek out those who might help me understand these complex river systems. Thus my travels led me to Berkeley, California, and the world-renowned geomorphologist, Luna B. Leopold (son of the famous environmentalist Aldo Leopold). Although Luna didn't know me, he was very accommodating and invited me to the University of California, telling me to bring my data and photos. In spite of my culture shock in coming from a cow town in Ennis, Montana, to Berkeley in the late 1960s, the many days and evenings spent with Luna set a new direction not only for my studies but also for the rest of my career. Luna became my mentor, over the years, and we eventually taught courses together two and three decades later. Luna's guiding philosophy was: The answer is not in the books or in the models, it is to be found in the river.

Luna also emphasized the need to document one's observations and, over time, to revisit the same rivers and review the data to understand response and channel change. It became critical to use rigor and consistency in field methods. My travels frequently found me in Pinedale, Wyoming, to review ongoing research of the U.S. Geological Survey and to observe and assist in sediment measurements conducted by Luna, William Emmett, and others on the East Fork River in the late 1970s.

Following Luna's recommendations, I set out on my new mission: to measure the water resource and the river's shape and structure before an activity, such as construction or logging, and then take the same measurements afterward for comparison. One of my first projects, Joe Wright Creek in the Arapaho/Roosevelt National Forest, ran close to the path of a proposed state highway construction project that threatened river stability and native Westslope cutthroat habitat. The environmental recommendations for the project would have cost the state additional dollars and thus were not implemented. But we were ready to document the changes after the highway was built.

We found that in the first year after the highway was built in 1976, there was a 200 percent increase in sediment load, and a doubling of the width/depth ratio of the channel—the river got wider and shallower. This change in the channel brought in deposits of fine sediment, potentially affecting fish habitat. From 1977 to 1979, an existing reservoir was enlarged on the same river. With help from the Colorado State University fisheries department, we found that sediment had increased by 250 percent and the native adult cutthroat trout population had decreased from 40 fish to 4. However, in the second year after construction, the juvenile fish population returned, and though the stream still had a lot of fine sediment, it was resilient and cleaned itself out within three years. Our findings became instrumental in helping the Forest Service and others emphasize how important it is to recommend ways to mitigate damage in similar work in the future.

Not all rivers recover as swiftly as Joe Wright Creek. Some, having suffered the effects of mining operations, have not recovered even after 50 to 80 years. I realized that there was a relationship between the character of a river and the speed with which it could fix itself. If the stream still has access to its floodplain and has good streamside vegetation, it can recover from disturbance much faster. On the other hand, some streams were very sensitive to disturbance and seemed to be severely impaired by practices that normally shouldn't have caused such a serious response. In other words, not all rivers respond similarly to the same use.

In my years on the river, I realized that the collected river data needed to be sorted by the type of stream surveyed. Just saying that data are from mountain or plains streams or from a gravel-bed stream was too broad to be of much help. This led to my 27-year development of a stream classification system.

(Often, government workers who try to collect information in the field ask me, "How did you get all of this data?" They tell me that they normally get caught up in the "administrivia" and when they start out the door to the river, they get redirected to "charm school" or some other conflicting, deskbound office assignment. I offer this secret technique to those inquiring:

Generally your boss wants you at the office *first thing* Monday morning, so you arrive at five o'clock on Monday morning and notice no one is there. Leave a note for your boss, saying, "I was here *first thing,* but looks like you're a little late coming into work this morning. I'll be collecting data up on Lick Creek and will see you later in the week." On Friday, you come in at 8 p.m. Again, no one is there, which allows you to efficiently take care of the "administrivia notes" left for you without distractions. You then leave a note for your boss: "It looks like you left work a little early today; sorry I missed you. See you *first thing* Monday morning." This can go on for a long time, and eventually they will not expect to see you in the office!)

My time wearing out waders in rivers was rewarded by a data set that was used to compare traditional river works with natural stable rivers. I could see that these river works, departing from the river's innate, stable form, led to great expense and great loss of habitat. These early observations motivated me to develop new solutions by breaking from tradition. The challenge was: How do I fix a damaged river, or even reverse the effects of river works?

With this in mind, my first river restoration project was in Montana in 1969–1970 near Big Timber, where I was assigned to restore Big Timber Creek above and below private land on Gallatin National Forest lands. A recent flood had scattered gravel and eroded banks, causing serious damage. To my dismay, a different federal agency had already been there and gone . . . and had "fixed" the private portion of the stream into a straightened, leveed channel. I then faced a real dilemma: How could I justify fixing the same stream twice in the same year, all with federal dollars? I knew the straightened channel could not be left like it was, as it would "head cut" up into the Forest Service reach, and sediment would pile up at the downstream reach. To restore the creek, I would have to gain the cooperation of the private landowner; a river does not respect property boundaries.

In late April, I was dropped off by helicopter on top of the Crazy Mountains, and I skied down to measure snow depth and water content every 500 feet in elevation. I then skied farther onto the private land where the straightened channel was located and met with Spike Van Cleave, the longtime owner of the ranch. I invited him to take a helicopter flight with me, since the copter was waiting below his ranch to pick me up. Never having been in a helicopter, Spike was more than excited to see his ranch from a perspective other than the saddle. I pointed out the "meander bends" of my ski tracks and asked him what he thought would have happened if I had not made any turns and had gone straight down the mountain. Flying over streams that had not been damaged by the flood, I pointed out the relatively predictable nature of their meandering character. Upon flying back over his straightened river, Van Cleave realized the mistake. He agreed to put the meander bends back on the straightened reach to be in balance with the reach above and below his place.

Despite some opposition from my own agency, which questioned why I would want to make a crooked stream, the meanders were appropriately restored to the stream on both private and public lands. The Gallatin Forest hydrologist, Leon Logan, then monitored this project from 1970 to 1972, documenting its success.

I often hear about the social and political constraints involved in river management and restoration. But these can be overcome. I recall a situation on the Roaring Fork River near Carbondale, Colorado, that could be titled "Against All Odds" or "The Need to Neighbor Up." About ten years ago, I received a call from the Department of Justice, which wanted to hire me as an expert witness for a Roaring Fork River dispute. Private landowners were pushing gravel around in the river and placing the gravel on streambanks for sacrificial riprap, all without a permit. This had been going on for several years, and as a result, huge fines were being levied and downstream landowners were suing the upstream landowners involved in the bulldozing. I pondered the situation, then agreed to be involved, but not in the form of litigation. I asked that they enlarge the earliest aerial photos (1936) and take new photos at the same scale for comparison. I stated that I needed three days to travel to the river and gather data, and then I would need all of the players in one large room. They claimed that due to the hostile nature of those involved, someone might get shot! Nonetheless, they agreed, and sure enough, everyone showed up.

Looking at the large audience, I sensed the open hostility but started out by saying that I was there to "represent the river." I proceeded to explain some fundamental principles about river behavior in layman's terms. Upon finishing, I asked how many people in the room would prefer to see the bank erosion reduced? All hands went up. Next question: How many people want to improve fish habitat? Again, all hands went up. Next question: How many here want to improve their land values? Again, most hands went up. My response to the group was, "Well I'll be darned, they told me that you all were at odds not only with the government but with each other, but I discovered you all have common objectives. You just don't have a common solution."

I then showed them the old photos compared with the new, indicating what had happened over time. A lot of the problems had come about by going against the grain, working against the central tendency of the river. I then put an overlay over the new photo, indicating what it would take to restore the Roaring Fork in these locations along the river. To make this happen, I told them, they needed to "neighbor up." This meant that everyone needed to work together toward a central goal and in a common direction to help the river, regardless of private land boundaries or affiliation. I also looked up at the Justice Department attorneys and asked them if they would be willing to drop all fines and litigation at once if all involved were to sign up to help restore the Roaring Fork. After a short break, they came back into the room and said yes.

A remarkable change came over the room when the attorneys said that they shared the same positive objectives for the river but that the last resort for them was litigation. The next step, I said, was to fire all of the attorneys in the room! Not because I didn't like attorneys, but because the money spent in litigation was not going to help the river and meet their objectives. That night they "neighbored up" and everyone, including the attorneys, was looking forward to a harmonious outcome. I developed the master plan for the river in those few days but declined to implement the work, as I did not want anyone to think the message was self-serving. I did, however, recommend others I knew who could put the design into action.

Today, knowledge of river management has improved remarkably. The public and agencies have a new awareness of the cumulative impacts of watershed development. People want their rivers back! Change can be a good thing, and when it comes to river systems, it is critical to change those variables responsible for continued impairment and to encourage recovery and escape from single-purpose river work objectives.

There's an old saying: "When you finally find an answer, they change the question." Well, the questions have sure enough changed. Now that we have learned from past mistakes, we are prevented from implementing new solutions by:

- **Tradition.** We need to break out of our comfort zones and take the risk of venturing forth to work with the river, not against it.
- **Technology transfer and education.** What has been learned about rivers must be shared with others. It's essential to learn and to improve the tools of the trade, including through formal training and mentoring. The standard methods manuals need to be updated and accepted standards must be documented.
- **Office modeling rather than field observations.** Although models can be helpful, if their assumptions and level of detail do not represent natural stable river systems, the results might look good on the beautiful graphic displays

> but not on the ground. It is essential to convert observations of river behavior into reproducible, consistent, quantitative expressions.

Because of available funding and public demands, we are currently in a rush to implement river restoration across the United States. In principle, this is a good thing; however, before you rush out to patch symptoms, you need to find the *cause* of the impairment. Restoration should not begin without the major critical element, which is a watershed and river stability assessment. This assessment identifies the location, extent, contributing land use, processes affected, and variables related to the cause of a river's impairment. It provides the inventory and analysis needed to understand and to be understood so that you can effect a positive change. The mitigation recommendation can often avoid an expensive direct modification to the system, and instead may suggest a more passive change in land use, depending on the recovery potential of the landscape or stream system.

I have implemented more than 200 river restorations over a 44-year period using natural channel design principles. Other practitioners have also instigated hundreds of projects over the past 10 to 15 years. Major restoration projects in Colorado on the Blue River, South Fork Platte River, Ohio Creek, and Little Snake River involved detailed monitoring by the U.S. Geological Survey (Lake Fork Gunnison River) and the Colorado State University engineering and fisheries departments, which documented the success of these large-scale river restoration projects. The lessons learned have contributed to the continued revisions and improvements of natural channel design.

As I look back over the numerous meandering trails I have left along the river—some for fishing, some for study—my reflections give me mixed feelings. I was glad to have been there, but sad that some of these trails continue to be too straight, too hard, too manufactured. What happened to the rivers we used to know? Have the lessons learned from the stark reality of irreversible change sunk into the heads and hearts of others? If we continue with the traditional methods we have forced upon our rivers, it will seal the fate for future rivers. We must work *with* the river, not *against* it. There must be a public outcry from those who recognize and see the change, but these voices must be heard over a great distance and received by those who can make a difference. Our voices have often been too faint. We must reach out strongly enough to shake the walls where lawmakers reside yet touch the senses of those who will champion the cause for the future.

The payoff for such efforts is an assurance that our children and our children's children can feel the magical attraction and love for being on the river and for fishing that was once afforded to me. As Luna Leopold often said, and as I tell my students now, "Protect the best, restore the rest."

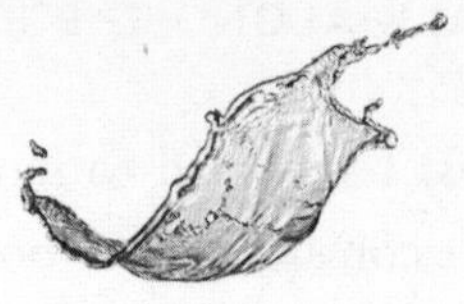

ON THE HEADS OF WOMEN

Kathy Robb

Kathy Robb became interested in environmental issues as a child, while riding in the backseat of the family station wagon out west. A partner with Hunton & Williams in New York, she focuses exclusively on environmental law, including litigation in federal district and appellate courts and advice on regulations, compliance, and environmental risks. She works on water issues under the Clean Water Act, the Endangered Species Act, the Natural Environmental Policy Act, water-related Superfund sites with PCBs and other contamination in sediments and groundwater, representing clients that include water districts, developers, electric utilities, energy companies, investors, lenders, chemical manufacturers, and paper mills. She founded and directs the Water Policy Institute, which seeks innovative, sustainable solutions to water supply and quality issues. Robb is co-founder and chair of the Women's Network for a Sustainable Future, a nonprofit organization advancing sustainability, and a board member of the Environmental Law Institute in Washington, D.C.

A GIRL AWAKES; washes her hands and face with warm, clean water running from the tap in her bathroom; dresses; grabs breakfast; brushes her teeth with cold, clean water from the same tap; flings her books and papers into a satchel that her mother worries is too heavy for her light and growing frame; and races out the door to the bus.

Another girl awakes, splashes cold water from a basin onto her face, grabs breakfast, and takes a large, empty container from the corner as she races out the door to join some other girls and start walking the six miles to her destination.

The first girl, my own daughter, is an American and lives in New York City. Every day she enjoys—without giving it a second thought, in the same way she breathes the air or assumes that she will have Internet access on demand—delicious water at the temperature of her choice from taps at home, at school, in museums, in public bathrooms, and in the pizza place around the corner.

The second girl lives in Africa but could be from any number of other places in the world today where there is no water service. She *is* the water service. And because she must walk more than six miles each way and stand in line for a couple of hours to bring water to her community, she does not go to school.

The first time we flush a toilet each day in the United States, we use about five gallons of water—more water than one out of every five people in the world has available to drink, cook, clean, and wash over a whole day.

There have been only a handful of instances when a new idea has marked unmistakable and instantaneous change from everything that has come before—where an innovation stands like a bright line between past and future and alters the way we think about ourselves and the world we live in. One such instance—at least for me—is that remarkable photograph of Earth taken from space back in 1972, during the last manned U.S. moon mission, Apollo 17. That iconic photograph, called the blue marble, demonstrates vividly that water (mostly salt water) covers about 70 percent of the planet. In fact, oceans make up more than 97 percent of the water on Earth, but it is expensive and highly energy intensive to take that salt water and make it usable. About 2.5 percent of the planet's water is fresh, but more than two-thirds of fresh water is locked up in the polar ice caps and glaciers. What is left, in aquifers, wells, rivers, and lakes, amounts to trillions of gallons, but still only a very small amount compared with all the water available. In 1972, when the "blue marble" photo was taken, there were about four billion people on Earth. Today, within just one generation, there are almost seven billion. It is projected that by 2050 there will be nine billion. Based on these statistics, water consumption is doubling every 20 years.

There is another reason that photo of Earth, looking like a blue marble floating in space, beautiful and fragile, is important. For the first time, we saw our planet as an indivisible whole, with no political boundaries and nothing to suggest that it is inhabited and being constantly changed by those who live here. The innovative technology that took us into space allowed us to see Earth from a perspective never before possible. That photo dramatically altered our appreciation of the natural environment and drove home that whatever happens on the planet happens to all of us. It illustrates that water is at once local and global.

I have always loved the ocean. My father had been in the Navy as a young man, and my brothers and I enjoyed playing in the waves in visits to sandy white Jones Beach, starting when we were babies. Later, growing up in Texas, I came to appreciate the difference between the ocean and a swim in a lake, and the contrast between the sweet-tasting water from the tap in New York and the municipal sources in Dallas, on the one hand, and the occasional well water we had when visiting friends in more remote parts of Texas. But it was only relatively recently, in Silicon Valley, that I came to understand the crisis in a lack of drinking water globally, and how it is tied to women and girls.

In 2000, journalist Ann Goodman contacted me with the kernel of an exciting idea about an organization for professional women, to promote and encourage sustainability. My enthusiasm and interest were immediate. I saw sustainability as a dynamic concept that, implemented in all its depth, could bridge differing views on environmental issues and bring together citizens' groups, regulators, and industry. Here, perhaps, was a way to move beyond the polarizing conflicts that had characterized the beginnings of environmental thinking in the United States. With a handful of similar-minded women, we founded the Women's Network for a Sustainable Future (WNSF).

At a WNSF-organized peer learning program in Silicon Valley shortly afterward, we heard from several technology companies about their efforts to bring laptops to the world's poorest children. Their goals were to teach and to connect isolated children to the larger world. Their stories were similar and fascinating. It turned out that to teach schoolchildren through computers, they first needed water.

The original intention of the companies was to distribute laptops and run educational programs through schools. They quickly found in their early planning, however, that in many locations there were no schools, or the schools were only sparsely attended, because there was no water or sanitary facilities available. Girls were especially affected by the lack of sanitation, because they required privacy, and because they stopped coming to school altogether once they started menstruating, because of the lack of separate sanitary facilities. Both boys and girls often missed school as a result of illnesses from contaminated water. And of course, many of the locations lacked the power that was required to run any laptop.

So these inventors of technology did what they do best—they innovated. They partnered with governments and nongovernmental organizations whose expertise was improving sanitation and water facilities, and with organizations whose mission was to help build schools. They provided the laptops and the educational programs as they had intended and were rewarded with the satisfaction of seeing the benefit to the children from their work.

Around the world—in remote parts of Africa, India, Brazil, Chile, Egypt, China, Russia, and Nepal, in Haiti and Togo—it is often women and girls who are responsible for providing water to households, women and children who are primarily responsible for collecting and managing water and making it safe to drink for their families. They travel several miles each day from their home in search of water, spending as much as eight hours a day collecting water. Every drop of water carried home must be managed carefully so there is enough for drinking, cooking, cleaning, bathing the family, and watering the vegetable garden. This means that millions of women and children spend hours each day searching for water, waiting in line for water, and carrying it back home on their heads, hips, or backs, causing damage and pain to their necks and spines. A water treatment engineer working in Haiti once reported watching women climb up and down a mountain path two miles each way every day carrying five-gallon buckets on their heads. Filled with water the buckets weighed about 40 pounds. The engineer scarcely washed for four days, to limit as much as possible any contribution to the women's burden.

Fetching water far from home can also be more immediately dangerous. Traveling out of their communities across comparatively long distances on foot, women and girls face the threat of sexual attack. In some cultures, rape victims, and the children of rape, are ostracized by the community. And I should mention that spending 60 percent or more of each day providing water leaves little time for other activities, including going to school, growing and preparing food, and working to produce income. It is estimated that in India alone, women spend 150 million workdays per year fetching water, equal to a national loss of income of 10 billion rupees ($217,000,000).

The issues of clean water and sanitation cannot be separated. Rural water sources in developing countries are frequently contaminated, and even if there is a more healthful source, added distance, fear of travel, and time constraints may result in women's obtaining, or accepting, lower quality water—not only in rural settings, but also in cities, where clean water may be priced out of reach for the poor or otherwise unavailable; indeed, contaminated water is the only option for some. Women and children are also the ones who suffer from inadequate sanitation during childbirth as the result of a lack of clean water. In Tanzania, women report taking clean water as the most highly prized gift for new mothers. And it is women who care for family members with waterborne diseases, most often their children.

Moving water to communities by means other than on the backs of girls is an engineering problem that has been addressed successfully before. After all, the Roman Empire began construction of its amazing gravity-driven aqueduct system almost 2,300 years ago. By the time the system was completed, some 500 years later, Rome's 260 miles of water infrastructure were capable of delivering 85 million gallons of water a day to the one million citizens of the ancient city.

Most scholars agree that any solution to water availability must include community women in decision-making and water management to reach effective solutions. Studies in Asia and Africa suggest that women are not often part of water management organizations in the community, resulting in decisions that are not optimal. For example, establishing a water source on a main road close to home may address the issues of long-distance travel to obtain water and free women to pursue education and income-producing work, but the location in a public place may have other issues that are not addressed and that are crucial for these women, such as safety and modesty.

Not surprisingly, water engineering tends to emphasize providing water facilities, leaving the social issues to be sorted out over time. Women's involvement in water management can sometimes be seen as largely a household function centered on providing, managing, and safeguarding water for the family, although studies show that women are equally interested in exploring ways to be engaged in income-producing enterprises. Of course, improvements in water supply address both. A gender focus on water management would involve a reexamination of the social approaches and how they might differ for men and women.

There is also agreement that engineering solutions must be compatible with the culture and on a scale with the problem. In considering how to provide safe, sustainable supplies of drinking water and improved sanitation and hygiene, the challenge is not finding solutions—proven, effective, sustainable solutions abound that are simple and inexpensive and can be taken individually and collectively. The issue is awareness and implementation.

By tapping into creative, innovative thinking, we are finding less costly and more efficient ways to address water issues and get those girls to school. Small projects like establishing a water purification lab for a local hospital where more than half the patients were being treated for waterborne diseases; using chemicals available locally to introduce simple methods of water purification; supporting local efforts to design a water treatment center; designing and building simple latrines; introducing pump handles that are easy to use and maintain; constructing a water tank that can be hooked up to existing, unused supply pipes, rather than building an entirely new, more expensive system; planting trees to combat deforestation and improve watershed quality; building a well pump that runs on the power from children playing on swings and other playground equipment; erecting stone barriers to prevent runoff and filter water—all these actions and many others have been taken by communities to dramatically improve water supply and, as a consequence, the lives of their people.

The United States has an unparalleled opportunity to establish and implement a strong global water policy that benefits the needy, encourages sustainability, advances economies, and saves millions of children's lives. In his Inaugural Address, President Obama vowed to the world "to work alongside you to . . . let clean waters flow." We need to make good on this promise. And we need to encourage developing countries to promote sustainable water through their regulatory frameworks.

The great poet Horace, who enjoyed the water brought to his city by those Roman aqueducts built 2,300 years ago, said, "To have begun is half the job: be bold and be sensible." We have begun the job of getting clean water to all people who need it, by identifying the problem and the answer. Now is the time to be bold and sensible, and finish that job, creating solutions that don't bear the weight of water on women's heads.

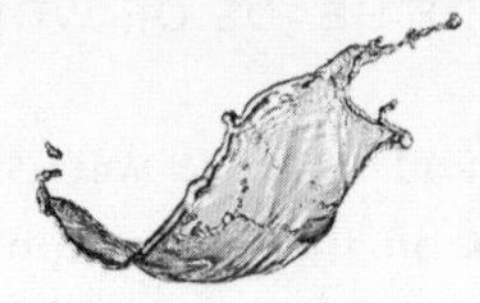

TWO NOBLE TRUTHS

Ashok Gadgil

Ashok Gadgil is a faculty senior scientist in the Environmental Energy Technology Division of Lawrence Berkeley National Laboratory. His research has concentrated on solar energy, energy efficiency, and indoor air pollution. In 1993, a deadly outbreak of cholera in his native India inspired Gadgil to focus his attention on waterborne diseases. He conceptualized an inexpensive water purification system that uses ultraviolet light and has been adopted by communities around the world. He has received numerous international awards, including the 2009 Heinz Award, which recognizes his work as a researcher, inventor, and humanitarian. He is currently working on affordably removing arsenic from Bangladesh drinking water. He also recently pioneered the design and development of a fuel-efficient stove to reduce violence against women refugees in Darfur as they collect firewood.

THE FURIOUS YOUNG schoolteacher stormed right up to me. My six-foot-one-inch frame loomed over his short and skinny physique, but his anger made me take a step back. I was shocked to discover it was directed at me! The year was 2000 and I was deep in rural Bangladesh to visit one of the severely arsenic-affected villages to see the tragedy firsthand. I hoped to understand the context of the problem and seek feasible solutions.

"You!" He fumed indignantly. "You have a return air ticket! You are here just to watch our misery and take photos," he almost shouted. "What choice do I have if the only water we get has high arsenic? I'm already showing signs of arsenic poisoning!" He put his arm next to mine. Our skin was the same color, but his was mottled with arsenic sores. "My future is limited, my health will soon fail, I will have trouble getting married, and won't live to see my children grow up. And you! You too started out from nearby Bombay, but you've managed to go to the U.S. Now, you'll soon fly back! What have I done to deserve being trapped here, like this?" I was taken aback at the outburst. He was right to be furious at this injustice of fate, and I was left without an adequate answer. There was nothing I could do for him, right then. All I could say was, "It could have been me in your place." In my mind, I resolved to do something about this; though nothing can be done about fate, surely science and innovation could at least outwit arsenic!

That year, 2000, the top one billion people in the world (including me) earned 78 percent of the world's income, and the bottom one billion people (including the Bangladeshi schoolteacher) earned 2 percent of the world's income. The disparity in research spending is most likely much larger; nonfatal problems of the top billion attract substantially more research funding than the life-threatening problems of the poorest billion—probably by many orders of magnitude.

My path to facing a furious, arsenic-scarred Bangladeshi schoolteacher had an unusual trajectory. In 1973, I came to the University of California at Berkeley from India to pursue a doctorate in physics, and a whole new world opened up. In India I was a classic math and physics "nerd," being thrilled by the Dedekind cut and special relativity in the eighth grade, delving into Richard Feynman lectures in high school, and getting a perfect 990 in the physics graduate record examination as a sophomore in college.

However, several events right after my arrival in Berkeley made me question the research trajectory that I had set on. In 1973, and again 1975, the United States was shaken by oil crises—for the first time, there were long lines of cars at gas stations. Suddenly the seemingly assured upward arc of American prosperity and indefinite economic growth seemed to falter.

Even so, America seemed shockingly affluent relative to India. I had always thought representations of U.S. lifestyles in Indian movie theaters were a romantic exaggeration, typical of Bollywood fare. On arrival, however, reality sank in: Everyone had a phone (back then, we Indians had to pay a hefty fee and wait ten years to get one), almost everyone had a car and a separate detached house (unthinkable for most Indians), and Americans were putting more fertilizer on their front lawns than India used for all its agriculture! When my American friends offered to drive me to see the nearby Oakland slums, I was even more amazed: The houses all had reliable electricity, 24/7 running water, flush toilets and phones, and paved roads in the front. "My" slum dwellers back in Bombay were living in rags, in illegally built bamboo-and-tarp shelters or under bridges, with no amenities.

This juxtaposition spurred me to start learning about India's poverty, developmental economics, industrial policy, and the political economy of development and underdevelopment. I also started wondering whether doing theoretical physics, with no practical applications in sight, would be the best way to help my countrymen. On top of that, in 1975, Indira Gandhi declared an internal "state of emergency" on questionable grounds. The government suspended basic human rights, censored the press, launched forced sterilization campaigns and jailed political dissidents. This made me question the relevance of my fascination with pure physics all the more. What did I really want to do? What was really worth doing? To which area of inquiry should I apply my mind?

Ultimately, I decided to transition my studies toward a more pragmatic applied physics. Soon after finishing my doctorate in physics my (Indian) wife and I returned to India with our eight-month-old daughter, with no plans to look back. I began doing research on energy efficiency and renewable energy at the then fledgling Tata Energy Research Institute. However, after five years of the company's rapid growth, I realized that my scientific aspirations no longer matched where it was headed. I finally made the difficult decision to leave India and return to Berkeley with my family, now with two daughters.

One of my first projects at Lawrence Berkeley National Lab was an attempt to address India's chronic battle with power shortages. Blackouts plagued most cities and industries, even as electricity was wasted in inefficient end-use technologies; lightbulbs are better at warming the air than lighting a room. The Indian power grid served many households whose lighting demands at night overwhelmed the system's capacity and led to blackouts. The majority of these households were also too poor to afford efficient compact fluorescent lamps (CFLs). It seemed a catch-22: The grid's capacity was inadequate to light up so many millions of inefficient bulbs. Meanwhile, these same millions of families were so poor they qualified for "lifeline" electricity rates that lost money for the power companies.

I realized that if the utilities could subsidize CFLs for their poorest customers, we'd have a win-win-win. I refined this idea and published several papers on its applications. In 1991, our attempts to implement the project in India succumbed to institutional politicking. Still, we persisted and succeeded brilliantly, first in Mexico, then Poland, and then in several other countries. My work with energy efficiency and lighting had a useful lesson: Lateral thinking can sometimes offer elegant solutions to complex problems.

Then I turned my attention to water. I had long noted the relatively high prevalence of waterborne diseases in India compared with their near absence in the United States. For that matter, my young daughters often suffered a bout or two of diarrhea every time my family visited India. I began to realize that although my generally healthy daughters bounced back, there were many children their age who never would. On one visit, my mother revealed that she had lost five nieces and nephews (cousins I never knew) to diarrheal death in their infancies.

Back in the 1990s, waterborne diarrheal diseases annually killed an estimated four million worldwide—mostly children under age five. About half the deaths were in South Asia. Infrastructure development and institutional motivations had failed India's children.

While pursuing other research threads, I started to collect and send relevant publications on drinking water disinfection to my scientist colleagues in India, suggesting they explore affordable and effective ways to produce drinking water from raw water. At first glance, UV disinfection appeared particularly promising. However, the reasons my colleagues offered for inaction were similar to mine: They were paid to do something else! They had to write the next journal paper or the next proposal, or review a manuscript, or whatever was the task du jour. The situation changed dramatically for me with the convergence of two events in 1993. Art Rosenfeld, my former thesis adviser, asked me to provide guidance to a mechanical engineering student who was conducting research for his master's thesis, and in exchange offered a summer salary for the student to work with me to assess the costs and effectiveness of drinking water disinfection with UV light. We discovered that for as little as just half a U.S. cent, we could disinfect more than 250 gallons of clear water with UV light. So, under ideal conditions, the *annual* cost of disinfecting one person's drinking water (two and a half gallons a day) would be less than two U.S. cents!

All that summer, I painstakingly incubated bacteria in water samples for my research on top of my home water heater. Meanwhile, another bacterium that was stoppable by my UV invention was barreling unchecked through South Asia. This outbreak of a mutant strain of "Bengal cholera" started in India and spread to Bangladesh and Thailand, killing tens of thousands of people month after month. I remember feeling horrified the first time I read the details of death by cholera. I was elated that I might have a solution to stop such an outbreak but devastated that this time around it was too late. I vowed then that I would pursue the UV water disinfection idea relentlessly until I got it right—it must function effectively, affordably, and robustly in remote parts of the developing world.

My early UV disinfection research was done at the Lawrence Berkeley National Laboratory—one of the most prestigious of the U.S. national laboratories. All of this work was initially done on my own time, time stolen from family and sleep. In the national labs, scientists are funded only by project; there is no core funding (or supported time) at the personal level to explore other research ideas, unlike for tenured faculty at a university. My colleagues generously helped by lending lab space and equipment, but I struggled for consumables and supplies.

In 1991, I won a Pew Scholars award for my work on energy-efficient lighting for poor communities in India. I was thrilled to discover that the award included a no-strings-attached research grant—which I immediately devoted to my UV disinfection work. Still, I knew I'd need more support. My initial requests to more than 50 foundations for funding were all refused or ignored, but in the end a couple of foundations agreed to provide modest support. Several UC Berkeley engineering grad students joined the project as volunteers.

My second lucky break was through Steve Witkowski, a program manager at the U.S. Agency for International Development, who saw the potential importance of a breakthrough in UV water disinfection. He defied institutional skeptics and naysayers to support my work with modest but steady funding. We held "stop cholera" workshops in Washington, D.C., and Bhubaneshwar in Orissa, India, to get feedback from a diverse audience about any potential drawbacks in our evolving design and to network with potential supporters. I was always a bit incredulous that such a simple, elegant design had not yet been tried; I worried that there was some fundamental flaw in our design that I had missed.

The principle behind UV Waterworks (the name of our UV disinfector) was to use a fail-safe technology that could provide clean water for the poorest communities in the developing world. This meant that it had to be highly effective, very low maintenance, of nominal cost, and robust enough to withstand harsh environments. This would take cycles of research, design, and testing—far beyond my personal capacity or finances. Just as this challenge seemed insurmountable, two dedicated researchers—Edas Kazakevicius, a young physicist from Lithuania, and David Greene, a recent graduate of Rice University—offered to move to Berkeley and work on this project full-time and for free. With this infusion of energy, we raced through design refinements and successive cycles of field testing, feedback, and redesign in India, Mexico, and South Africa.

As my employers, the University of California and Lawrence Berkeley National Lab own the patent for UV Waterworks; they licensed the technology to a California startup, WaterHealth

International, in 1996 and launched it into the real world. In that same year, this invention won the *Discover* award for the best environmental invention of the year and *Popular Science* magazine's Best of What's New award. Both awards brought a burst of publicity and enthusiasm for the newly formed WaterHealth.

One day, I was feeling a bit down, struggling with balancing home and work and fund-raising and research. I was pleasantly surprised when Natural Resources Defense Council's Peter Miller walked up to me and said "Great! You have such an energy-efficient water disinfector. Let's see how to get it tested in Africa!" Through him I learned of the Lily of the Valley hospice near Durban, South Africa, for abandoned HIV-positive infants, established by a retired Afrikaner farmer and his wife. Their only water supply was from a bore well that tapped a groundwater aquifer. Tragically, this water was badly contaminated from nearby pit toilets. About half the infants admitted to the hospice emerged HIV free and were moved to orphanages. The other half were loved and cared for until they died. Heartbreaking though it was, I realized there was still something I could do. We installed a UV Waterworks unit at the hospice to provide safe drinking water for the infants. Suddenly, diarrheal disease declined precipitously, offering a better quality of life even for their brief childhoods.

Each village now has its own safe drinking water supply in the form of a micro water utility. The water treatment technology is housed in a small structure on community land near the village's original raw water supply (often a local pond or river). These WaterHealth centers collaborate with local nonprofit organizations to provide essential community education in public health and hygiene, financing from a bank, construction, operation, and maintenance. The village council (a locally elected body) owns the center. The costs are recovered by selling the drinking water to the local community. Prices vary depending on times and places, but affordability is one of the key tenets of the model. In 2007, the prices in eastern Andhra Pradesh in India were about two cents per two and a half gallons of potable water, 140 times cheaper than the local price of bottled drinking water.

In 2005, I visited India's first fully functioning Center in Bomminampadu, a remote village outside Vijayawada, in eastern Andhra Pradesh. We had just hired a thin, quiet man named Venkanna as our center technician. Up till then, he had been struggling to find work and was on the verge of leaving the village to expand his search. He told me this job had been a lifeline; he had been deeply unhappy at the thought of leaving behind his family and community to find work. "This job has changed my entire outlook!" he crowed. I met Venkanna—now a beaming and confident man—again in 2008.

By the end of 2008, 300 of these micro water utilities were either in operation or at various stages of construction, with a service capacity for more than a million villagers. They employed several hundred people in India and served safe drinking water to more than a million rural villagers. I was visiting one of "our" villages when a mother thrust her chubby infant into my arms. Looking down at his smiling face, I realized that what had been just a spark in my mind way back in 1993 had come a long way.

Though UV Waterworks has been demonstrably effective in addressing some waterborne illness in the developing world, there are other problems beyond its scope. The horrific arsenic contamination in Bangladesh, for instance, has been well recognized for more than a decade, but we've made little progress in reducing its impact. More than 50 million people are affected by arsenicosis in Bangladesh and the neighboring Indian state of West Bengal, in what is rightly called the largest case of mass poisoning in history.

That encounter with the furious schoolteacher remains with me even today; since then I've invented two novel methods of arsenic remediation—both are technically effective, robust, and affordable but are still in early stages of field testing.

On one site visit to West Bengal, I decided to tour some unsuccessful projects to remove arsenic from water to learn what mistakes to avoid. In every case, the science had not failed, but the scientists had. They had overlooked the needs of the community and set up technologies that were too expensive, complex, or delicate for the rural environment. We faced a harrowing scene: Many villagers in Murshidabad district had had limbs amputated as the result of arsenic-induced gangrene; family members had died from arsenic-induced cancers; and many of the rest had spiraled from poverty to destitution, losing all hope. In reality, much of this suffering might have been prevented, and still can.

That angry schoolteacher I'd met in 2000 was right to be angry. Arsenic in groundwater is an act of nature, but its presence in drinking water is an act of human society. Our collective failures in addressing arsenic in groundwater are not an accident or some act of fate. They reflect an uncomfortable truth about institutional priorities, but simultaneously reflect an opportunity to change the world for the better. My way of making a difference has been through science, and two additional noble truths sustain me in my work. One is to always remember my humanity, and the second is to always move forward with optimism. The humanity makes me care, and the optimism makes me keep trying.

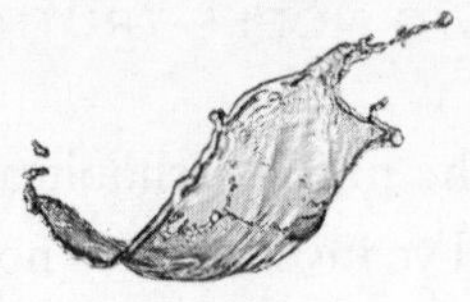

NATURE ABHORS A MONOPOLY

James Workman

James Workman, a Yale University graduate, is an award-winning journalist, a speaker, and an entrepreneur. He has prepared speeches for statesmen from U.S. Interior Secretary Bruce Babbitt to South Africa's Nelson Mandela and led a campaign that blew up obsolete dams to replenish dying rivers. For seven years in Africa and Asia he helped forge the landmark report of the World Commission on Dams; led radio and TV documentary research safaris; and advised global businesses, aid agencies, and conservation organizations on water policy. Workman is the author of the critically acclaimed *Heart of Dryness: How the Last Bushmen Can Help Us Endure the Coming Age of Permanent Drought,* published by Walker in August 2009, from which sections in this essay have been adapted. He lives in San Francisco, where he has co-founded new ventures for trading the human right to water and energy, through SmartMarkets, Incorporated.

WHEN I WAS A CHILD, I played as a child, and what I played was Monopoly.

On cold, dreary nights nothing so invigorated the spirit as the chance to impoverish my family. I'd later earn a D in college microeconomics, but I caught on quickly to the game's cheerful goal: domination, exploitation, and enslavement of others. Long before Gordon Gekko or Ivan Boesky, Monopoly revealed self-interest as virtue. Adam Smith's "invisible hand" guided our metal race car tokens around colorful real estate squares. Less cunning players lost fortunes to luxury taxes and jail and to my big red hotels on Park Place and Boardwalk, and I laughed while driving rivals into bankruptcy. Monopoly reduced war and violence into the peaceful art of making money. And yet, for all its educational drama and avarice, one option in the game never squared with the abstract ideal of unbridled free market capitalism: Water Works.

On that square a disembodied spigot protrudes from an invisible wall. No color. No flair. Nothing left to the imagination. Even my modest home had several taps; where was the thrill in acquiring another? More to the point, this "utility" was unprofitable; after paying $150 for the title, you couldn't develop on it, and only got a $75 mortgage. Rental, limited to four times the roll of the dice (average $12 to $16), could never pay off big. Against high-opportunity costs of Water Works, savvy players seeking a higher return on investment held out for Marvin Gardens.

Invented during the Depression by Charles B. Darrow, Monopoly has aged far better than I have. Its timelessness lies in its ability to tap two of the three instincts that guide human behavior: greed and fear. The third driver kicked in during late adolescence, distracting my attention span, so when I became a man, I put away childish things. Only then, out in the real world, did I discover that the true path to absolute monopoly lurked in the shadows of the least glamorous commodity.

As any antitrust lawyer will confirm, the definition of "monopoly" is slippery and elusive. Even investors in the game's real-life inspiration, Atlantic City, reveal how a magnate like Donald Trump could never control the will of the people. Cornelius Vanderbilt and J. P. Morgan profited at the public's expense but created enough new wealth for competition to dilute their monopolies. AOLTimeWarnerCNN is still recovering from the burden of its monopolistic "synergies." Busting up AT&T left it stronger than as a trust; Microsoft bullies have ceded ground to the current "monopoly," Google. As for John D. Rockefeller, Andrew

Carnegie, and Cecil Rhodes, we can walk away from oil, steel, and diamonds—luxuries competing with cheaper alternatives. So really, is there any such thing as a long-lasting, people-enslaving, all-powerful monopoly?

Oh, yes.

The insidious thing is that few people are aware that it exists, even as it manipulates us all. The ultimate monopoly turns out to be quiet and faceless. It bears no high-profile brand name. It rewards no celebrity CEO, launches no marketing campaigns with catchy jingles or witty ads. And yet it controls every meaningful aspect of our existence: how we live, where we live, and in some cases *if* we live.

Odds are that you and I are beholden to this monopoly for our health, energy, food, and income. Never mind Keanu Reeves science fiction movies; our bodies and souls are enmeshed in the Matrix of Life, and access to the supply, demand, quantity, and quality of that matrix sits in the hands of a few strangers operating a quiet, bland, yet monolithic entity: Water Works.

My epiphany about the profound extent and sinister nature of this water monopoly did not come through grad school econometrics (remember my D?) but through a harrowing humiliation in the Kalahari Desert of Botswana. There, forgetting all I'd learned in Monopoly, I lapsed into an altruistic impulse and intervened on behalf of the last free bands of indigenous people, known as the Bushmen.

In the austral summer of 2002, to force Bushmen to relocate outside the Kalahari, the government of Botswana abruptly cut off its deliveries of fresh water to them and destroyed their water pumps and storage.

Weeks after reading about the water cutoffs, and the ensuing siege, I arrived in Botswana driving my shiny, red eight-cylinder

Land Rover and met with and interviewed some human rights activists outside the Kalahari. They had filed a lawsuit but believed their case would be stronger if they could show defiance by those who voted with their feet. Overcoming my reluctance (and fear), I was eventually persuaded by them to pose as a tourist, detour from the wildlife-viewing areas, and sneak in provisions to bands of Bushmen who decided to stay put.

The next day I set out for Vulture Water, a translation of the Ganakwe name Metsiamenong. The four-lane paved highway narrowed to a one-lane road, which soon turned into deep, soft, hot, miserable sand. Fifth-gear cruising dropped to second-gear grinding, but by evening I had entered another world.

At dusk I lurched past huge, beautiful, black-maned lions on the prowl. But I resisted photographs; I was in too much of a hurry to "save" the people celebrated by *National Geographic* or the movie *The Gods Must Be Crazy.* Recently, geneticists traced markers in the Bushmen's mitochondrial and chromosomal DNA that confirmed their status as the nearest blue-blood relatives to our common anthropological Adam and Eve. Most had assimilated into the fringes of agricultural and industrial society, but not yet these defiant "Last of the First."

The next day, during my one-man water smuggling operation, I paused in the late morning heat to congratulate myself. Here I was, a recognized water resource management expert, carrying contraband to the most marginalized inhabitants of the most marginalized continent. As I devoured lunch and chugged canteen water, I jotted in my journal, "I feel this is the most important thing I have ever done in my life."

The ancient Greeks knew such self-righteous babble as hubris, a form of willful pride based on ignorance of reality. Something the gods loved to punish.

The torque and technology of Land Rover insulated me from my hostile surroundings until the epicenter of the subcontinent, the point farthest from any reliable spring or river. I'll never know when the air intake hose jiggled loose and began sucking unfiltered air into the carburetor, but I know exactly when gas vapors and dust accumulated to form an oxygen-blocking clog.

All at once the engine wheezed and died. I soon appreciated that I lacked the right tools and experience to open it up. After my screams tapered off, the insect noises of the arid Kalahari rose in the silence. Suddenly, all that modern internal combustion technology and gasoline fuel was worth far less than my precious containers of water.

I had driven in too far to walk back out. Moisture was fleeing my body by the second. I recalled those lions, and imagined the hyenas that inevitably followed, and what my bones might sound like when splintered to dust in their powerful jaws. In the dehydrating heat I began to panic. The worst thing I could have done was seek comfort in booze, so naturally I began chugging wine right out of the bottle. It didn't improve my judgment that night.

What did help, the next morning, were the Kalahari's amoral surroundings. To paraphrase Samuel Johnson, when a man in a desert knows he will run out of water in a fortnight, it concentrates the mind wonderfully. As I grew accustomed to the Kalahari, I realized all life out there thrived on a vibrant, fiercely competitive economy of water.

Camel thorn *(Acacia erioloba)* roots seek embedded moisture beneath its shade. Shrubs grab the wet molecules in tiny pockets of sand, where leaf patterns funnel scarce raindrops. From such plants the beetles and blind mole rats and springbok captured their moisture, and the blood of these small prey species slaked the thirst/hunger of barking geckos and snakes and birds and leopards. Even "pure insectivores" like yellow mongooses, or "pure carnivores" like cheetahs, skipped the middlemen to devour moisture-rich succulents and *tsama* melons. These dynamic exchanges reminded me that life wasn't merely linked through a food cycle, so much as a *water* cycle. And at the center of this tight economy, for at least the past 30,000 years, thrived the Bushmen.

What's more, I saw that no single organism dominated all. Nature abhors a monopoly.

Hearing a low rumble in the distance one night, I chased down headlights on foot to catch a convoy of Afrikaner tourists heading in a different direction to wildlife pans. They were far better equipped and more knowledgeable than I, and with the right tools and savvy opened up one end of the valves, cupped the air intake, and sucked out the clog. My engine roared and, as a measure of my gratitude, I gave them the rest of my wine.

"Ag, you're a lucky man," one told me. "Traveling alone like this."

"You mean *stupid* man. I know. I owe you more than you can imagine."

"If you don't mind our asking, um, what the hell brought an American out here headed off in this direction?"

I explained, sheepishly, that I was researching the causes and consequences of water scarcity, and what happened when society ran out of water.

"You know," he deadpanned. "There might be easier ways to find out than this."

I nodded.

"You ought to come follow us. Stick with our convoy."

His offer was tempting. It would have offered safety in numbers, relaxation of wildlife watching, and companionship in a lonely landscape. Only a delusional fool would have politely declined.

I politely declined. But in my defense, over the past 24 hours I had undergone the equivalent of Saul's conversion; on the road to Metsiamenong I was no "water management expert," a title for those with an abundant supply of water to manage from an office on high. Instead, water was vanishing everywhere. Snowpack and glaciers melted from the Rockies to the Himalaya. Reservoirs behind 800,000 dams evaporated at a faster rate each day. Human thirst multiplied exponentially from economic and population growth. I could delay the reckoning with scarcity but never escape it. No one could. Against the menace of radical climate flux, we would have to live with less, a disciplined water economy known only to those Kalahari plants and animals and people.

Clearly, Bushmen didn't need to be "rescued" by a bumbling naive California idiot. But that idiot, his arid home state, and the rest of the increasingly hot, thirsty world sure needed Bushmen to help us survive the Dry Age of our own making.

So I waved off the kind and heroic tourists and rolled on in the opposite direction. I had glimpsed the potential horror at the end of water, a horror the dissidents out here embraced by choice. Driven by a new urgency—self-interest, not altruism—I sought to absorb Bushmen coping mechanisms. Nervously checking my engine air hose every hour, I lurched over the maddening sands into the arid savanna, where man was born and where his scruffy unwashed descendants still gave birth and teased each other and smoked and hunted and danced, and died.

Hours after my engine recovered, I somehow stumbled my way into the Bushmen camp at Metsiamenong. Small groups of people were congregated around tiny fires. They smelled of wood smoke and stale sweat, and it appeared they had never washed the only items of clothes they wore. But no one here appeared overly concerned or panicky; no one seemed a nervous wreck on the verge of collapse. I mean, besides myself.

Children played. Men stirred embers. Women laughed at bawdy jokes. I imagined this was how it had been before Botswana had entered their world with the intent of improving their lives by extending its power through water deliveries. Now that existence had returned after Botswana changed its mind. Officials had welded shut the only well in the Kalahari, destroyed whatever storage they could find, and laid siege.

The band received my smuggled provisions with gratitude, but then set them aside. What they really thirsted for was news from the outside. What was Botswana doing? Why did it behave this way? Was it the same in my country, where I came from?

I had to explain, through the broken English of a young man I found there, that in the world outside the Central Kalahari, Botswana's government took immense pride in how it emulated—and in some cases surpassed—the United States, particularly when it came to resource management. It valued precious hardrock minerals more than water, just like the United States. It displaced wildlife with subsidized cattle, just like the United States. And it emptied its nature reserves to provide pleasure grounds of the rich, just like the United States. Days earlier, Botswana had defended its water cutoff policy and siege, after the model set by the United States: Displace indigenous families who got in the way of the national interest.

Doing so had made Botswana into the donor darling of Africa and helped the country graduate over a few short decades from the second poorest to the fastest growing and now middle-income nation. Outwardly, the oldest democracy in Africa peacefully embraced markets, shopping malls, and property rights. Beneath it all, however, the country quietly anchored the Earth's century-old diamond cartel, De Beers, which in turn was subject to Botswana's most powerful instrument of state coercion, a water monopoly.

The "Water Works" monopoly can make or break both property rights and human rights, and operates at three levels: federal, state, and local. First, the federal monopoly constitutionally binds every raindrop in its borders under a single entity. True, this entity can create friction in the 263 cases where fugitive water flows across borders, like between the United States and Mexico or Namibia and Botswana. But throughout history competing monopolists avoid violence, finding it cheaper and safer to collude over water use within their own borders.

Second, the federal monopoly controls states. On the Colorado River, seven entities north of the Mexican border—Wyoming, Colorado, Utah, New Mexico, Arizona, Nevada and California—are granted permission to use federal water in their own monopolies. Likewise, it divvies up the Flint River among Georgia, Alabama, and Florida. And so on.

Last, each state in turn dictates which businesses and cities get how much water for what purposes. Farmers won agricultural monopolies, followed by industry, leaving 97 percent of the population utterly dependent on total monopolies known as municipal water districts.

The hallmark of monopoly is absence of choice. Status aside, when it comes to water, the almighty consumer is, economically speaking, impotent. No competition, no incentives to improve. You can't wake up and say, "Honey, this water tastes rusty, gives the kids a skin rash, and kills the begonias. Maybe we should switch to another utility." Nor can you get the bill and say, "Whoa. They're charging seven dollars a month for us to hose down the driveway, take 45-minute showers, fill our cracked swimming pool, and soak the lawn at midday? That's outrageous. Let's see if we can negotiate a better price."

Most monopolies are public, but for roughly 15 percent of us, the utility is private. That simple distinction has fueled the flames of ideological strife.

If it's public, say conservatives, a monopoly's interest is to keep rates obscenely low to curry favor with voters, causing waste and bloat. Democracy rules. But it tends to cut short the career of a fiscally sound green politician, of either party, who campaigns on a platform to "slash subsidies that keep grocery store meat and produce too cheap; end irrigated crops grown in the arid West; curb lazy bathing habits that are depleting rivers; ban flush toilets as obscene anachronisms that waste drinking water and precious fertilizer," but who adds, "If you vote for me I'll quintuple your water bill."

If utilities are private, warn liberals, the monopoly could fall into the hands of Parisians or Berliners. French creativity or German efficiency aside, the transatlantic distance scares Americans. People understandably fear having their lifeline operated by executives beholden to day-trading shareholders who can dump stock in the company, leaving local ratepayers stuck where they are with what they've got. What's more, private monopolies face two demands public utilities escape: profits and taxes. The pressure to pay both skews delivery toward affluent apartments over the rural poor.

Whether private or public, however, the monopoly locks five billion humans under the ironclad rule of an uncompetitive Water Works. Even the 1.1 billion with no improved water services face an informal feudal monopoly imposed by a village leader, tribal chief, or warlord who controls access to "his" spring, stream, pump, or well. Like a petty version of an Egyptian pharaoh or a Roman Caesar, he controls the 72 percent of human bodies that consists of water. Just try to negotiate with that.

The most troubling aspect of this arrangement comes when authorities explain that total dependence on a water monopoly is perfectly, inescapably "natural."

The cost of a single water monopoly is lower (due to economies of scale) than if two or more entities tried to compete for our thirst. Think about it: all those dams, diversions, dikes, pumps, canals, aqueducts, sewage treatment and water purification plants, PVC pipes and meters running to your toilet and sink. Now, imagine doubling them. Impossible. And so there we are, locked within our self-imposed "natural" prison. Even as supplies grow scarce under the onslaught of a warming planet, we force ourselves into an ever deeper vicious cycle of increasing thirst, helpless dependence, and addiction to wasting the most precious resource on Earth.

Perhaps there's hope. As I arrived and made camp in the Kalahari and walked and ate alongside the Bushmen, it struck me that they may be among the only people on the face of the Earth who still remained liberated from the shackles of monopoly. Somehow they thrived, despite harsher conditions than our own. If they could find voluntary, collaborative ways to squeeze every drop and make it endure, why not us?

Of course, even the Bushmen could not escape the dictatorship of nature. No one is free from that rule. We can push back time, borrow against the future. But nature always has the last word. We exist in a state of comfortable naïveté, insulated from nature's jagged edge—then we feel the sting of reality. A Land Rover engine freezes. A levee breaks above New Orleans. California's eighth largest economy wilts under drought. Atlanta—the fastest growing city in the United States—had only 60 days of stored supply before millions ran out of water. Virginia and Maryland fight over the waters of the Potomac for the first time since the Civil War. China lacks enough water to feed itself. Grain prices double. India's farmers, deprived of groundwater to pump, sell wives into slavery to pay off their debts.

What's more, I saw Bushmen were not superhuman. They could grow as irritable and suspicious as anyone. They felt jealousy and hunger pangs and thirst, just like us. They generated conflict over scarcity. But somehow they managed to channel their emotions and self-interest into a self-regulating interdependence. They transformed tension into laughter. And water, however scarce, remained their cohesive social glue.

Over the next five years, returning into and back out of the Kalahari, and poring over every study I could find, I tried hard to grasp how they thrived. To me their approach boils down to "Seven habits of highly effective water users." Bushmen turn scarcity into relative abundance as they:

7. *Seek shelter*. On an increasingly hot planet, the only enemy is the sun, a force of death, like the hyena. Where the rest of us might see a water storage reservoir, Bushmen see a foolish sacrifice to the sun gods. They trap and secure water in closed, concealed, and buried storage and food, to block evaporation—a strategy the industrial world is only beginning to adopt, through aquifer recharge.
6. *Consume local*. As Bushmen eat and drink closest to the source of water, they avoid leaks, spills, pollution, and exposure in transport. Bushmen find the shortest distance from the ground to the hand to the mouth; Botswana and America convey water across continents to produce something, then ship it back, sacrificing billions of gallons en route.
5. *Diversify supply*. Bushmen obtain 80 percent of their moisture budget through dozens of plants; the rest they capture from pockets of water in trees, sip wells, pans, and hunted

animals. The rest of us depend on one dam, which can fail or be depleted. To spread out risk and secure resilience, we may learn, like Bushmen, to tap into dispersed natural infrastructure like groundwater, fog, dew, reuse, runoff, or rainwater tanks.

4. *Devolve decisions*. Bushmen practice the oldest form of self-governance but don't elect or submit to a coercive authority; a dry economy can't afford police. Instead, responsibility is dispersed to each individual. Trust emerges through voluntary interaction. Group decisions—on hunting eland, gathering *bi* (milkplant root), trapping *kori bustard* (a terrestrial bird), gathering tsama melon, inhaling from a sip well—follow the basis of logic and experience. This self-regulation rewards "water management experts."
3. *Own shares*. Libertarians note: "No one washes a rental car." But libertarians wash cars they own because of drinking water they rent cheaply or for free. They may hold the deed to land, house, and furniture, but they don't own water flowing through the pipes. Yet Bushmen can and do hold in trust a Kalahari pan; a cluster of shepherd's trees; a wild fruit or gathering ground; a hunting area; ostrich shell canteens; arrows that mark a kill. All these convey informal title, and Bushmen, like all of us, look after what they own. That care leaves enough to ensure an unspoken right to water; no one who's thirsty is denied enough to drink.
2. *Encourage trade*. Bushmen are not "proto-Marxists," and would most likely nod to Adam Smith. They truck, barter, and exchange water resources in a constant, complex, and transparent network while the community tracks and exerts peer pressure on those who can reduce demand, thus increasing collective supply. The band rewards frugal, resourceful, innovative, and efficient use with esteem. By contrast, water monopolies encourage excess; punish efficiency with unfair rations and

higher fees; and generate conflict, as every drop you save your profligate neighbor can use.

1. *Unlock monopolies.* This is the crux: liberty to choose. When the government of Botswana cut off its water deliveries to Bushmen, it unwittingly thrust them into a state of freedom. Monopolies become insidious by fostering addiction to their single entity. As one Bushmen matriarch, Qoroxloo, put it, "When I was young the men hunted and we got our water from the roots of plants. We lived well, and people only died of old age, not of diseases. But then government officials started bringing us water . . . and now that I think about it, I believe it was to make us dependent, and abandon our traditional ways, and get us to move." You cannot live free when you depend on one entity for all water without which you die. Only if we loosen the rigid forces of monopoly can we negotiate the terms by which we choose to live.

I was slow to absorb all this. During the second of my journeys into the Kalahari, I moved with tape recorder and camera from one Bushman to the next until coming to that matriarch, Qoroxloo, hoping to extract Important Answers to Profound Questions, such as: "What will you do without government water?"

She kept scooping flesh out of a tsama melon, trading gossip with another.

"How are you going to manage water during the drought?"

The old woman shrugged without looking up and shifted back on her heels. Next to her a small fire burned. It was more smoke than flame, but never seemed to go out.

I persisted. "Do you think you could manage enough water for your family and your band to last until the rainy season?"

Like others before her, she grew evasive. Repeating the question through my translator met with silence. At that time, her caginess didn't make sense. Years later, it has begun to. It wasn't that Bushmen didn't want to answer such questions; they just couldn't. My grilling a Bushman about how humans must manage water was like a Vatican cleric interrogating Galileo about how the sun must orbit the Earth.

To adapt to water scarcity we need not abandon eBay or Amazon, SkyMiles or Facebook. To the contrary, those Web 2.0 technologies can help us unlock incentives, track use, earn trust, accrue credit, enable exchanges, and break up the water trusts by creating virtual click markets within brick-and-mortar monopolies. We can keep the physical infrastructure; at the same time, we can own an equal "share" of water to use as we please. If we use it all, we owe nothing. If we use more, we pay dearly. If we use less, we can save, donate, or exchange our savings with others, rewarding efficiency and triggering a collective race to conserve.

The Bushman code of conduct may help us escape a Hobbesian or neo-Malthusian nightmare. Prepared for extreme deprivation, Kalahari Bushmen chose the hard but free responsibility of a dry reality over a government-dependent monopoly that bred a false sense of water abundance. Outside the Kalahari reserve the so-called civilized world found that for all our military might and Internet bandwidth, certain things still lie beyond out grasp. We discover we cannot "regulate" barren rivers and depleted aquifers any more than we can "regulate" our climate, clouds, or rain. Out here, while elected leaders kneel to pray for rain to replenish their government monopoly, the increasingly dry, hot wind whistles through the thorn trees in the central Kalahari and whispers the ancient secret that those last defiant Bushmen never forgot.

We don't govern water.

Water governs us.

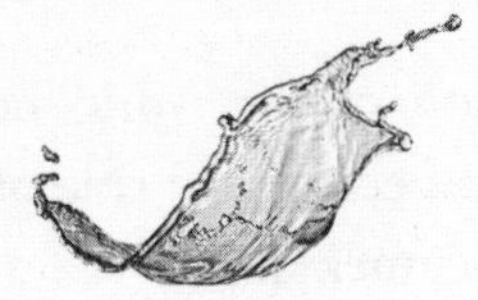

WATER IS LIFE

Alexandra Cousteau

Her legendary grandfather, Jacques, and father, Philippe, explored the ocean's wonders, and today social environmental advocate Alexandra Cousteau looks for new ways to conserve marine and freshwater ecosystems. Advocating conservation and sustainable management of marine resources for a healthy planet and productive societies, she works to inspire and empower individuals to protect not only oceans and marine creatures, but also human communities for whom fresh water is essential.

WHEN MY FATHER, Philippe, was four years old, his father, Jacques-Yves Cousteau, taught him to dive. My grandfather gently suited up my father's small frame with a mask, tank, and fins, and then took his young son's hand in his own and led him into a virtually unexplored world that teemed with life. My father was so exuberant about all that he saw beneath the calm surface of the water—a darting school of fish here, a brightly colored coral there, a waving forest of life just beyond—that he repeatedly tried to call out to my grandfather. He was blissfully unaware that each exclamation caused the regulator to fall out of his mouth, which my grandfather deftly and repeatedly replaced to keep his small, excited son from drowning.

When they finally got back aboard the ship, my grandfather admonished my father for his reckless enthusiasm saying, "You must be quiet underwater because it is a silent world." My grandfather's description of the world he had introduced my father to that day later became the title of his best-selling book and Oscar-winning documentary.

At the time of my father's first dive, the ocean was still a mystery to people. For thousands of years, it had been seen as an immense, unknowable universe in which the magnificent variety of life was hidden from sight. In that way, it was indeed a "silent world." Much of what we know about the oceans has been discovered in just the past century. Before that, the oceans' depths were mostly a mysterious, dark emptiness that our ancestors filled with monsters, from the giant squid to vengeful gods, who were bent on carelessly casting their boats aside. The imaginations of cartographers and the inkwells of poets shaped a view that, though exciting, was almost entirely wrong. It's easy today to excuse their notions as simple foolishness, for they had never dropped into her depths or seen for themselves the bounty and beauty of her secrets. But we have. And sadly, it is a willful and inexcusable form of ignorance that marks so many of the decades since my father and grandfather joined other pioneers in exploring the depths and pathways of our planet's water. We have denigrated the very oceans that cover a full 70 percent of our home, and we have ignored and at times vilified the inhabitants, with whom our distant ancestors used to dwell.

Forty years after teaching my father to dive, my grandfather took me on my first visit beneath the surface of the ocean in the same area of the Mediterranean, off the coast of southern France. It was 1983, I was seven years old and 40 years had passed since my father's first dive. Our experiences were quite similar in many ways. I stood weighed down by heavy equipment and concerned about how I was going to breathe in this strange, new world I was about to visit. I took a few tentative breaths and then plunged in—my worries quickly vanishing into the shimmering constellation of tiny, silver fish that enveloped me into lively formation as though I were one of their own. The light and the grace of this mass transfixed me—thousands of individuals moving both independently and in perfect unison through an entirely new world. The spell they cast that day has yet to wear off, and I still long for the moments when I can slip beneath the waves to experience their extraordinary and captivating universe, for I have seen its indescribable beauty.

But I have also seen firsthand the consequences of carelessness and a lack of understanding about our water planet. Sadly, though my father and I shared the same feeling of wonder for what we first saw beneath the waves, I stepped into a very different ocean than he had. The giant grouper my father had observed, weighing as much as 400 pounds, had been wiped out. Today, the red tuna that I observed on my first dive are quickly disappearing, and by the time I have children old enough to dive, this species of tuna will most likely be extinct.

Given the multitude of threats facing our oceans—from overfishing to ocean acidification to coastal development and so many critical issues in between—there seem enough problems to spend several lifetimes trying to solve just a few of them. But like my father and grandfather, I find myself drawn to the larger subject and searching for new ways to address the great challenge before us: exploring new ways of understanding what it means to live on this water planet in a time of growing demand and changing climate, and engaging people in meaningful dialogue about how we will shape the future our children will inhabit.

We live on a water planet. This transparent liquid is the foundation of all life on Earth. It ensures our survival, fosters our development, and enriches the cultures of our civilizations.

Yet despite the best efforts of scientists, filmmakers, and explorers, like my father and grandfather, our generation knows little more than theirs did about the ocean depths or the fragile scarcity of our freshwater resources. And while we debate the challenges climate change will bring to our lives, we have scarcely begun to recognize that *water will be the vehicle through which climate change is felt.*

Water is Earth's great storyteller. It is the mark of sustainability in a society and the canvas on which the pendulum of our struggle to maintain balance traces shaky arcs that ripple throughout our systems. And it is within these telling ripples—the shrinking surfaces of our ice stores, the erratic runnings of our rivers, the shifting patterns of precipitation, and rising of our seas—that we'll feel the effects of climate change first. And in the face of such a challenge, in just such a moment, we can't afford to make the focus of our conversation on water just about the fragility of coral reefs, the scarcity of free-flowing river habitat, or the depletion of fish stocks. We can't continue to divide over protecting fresh water or focusing on the world's oceans. We have to return to the simple truths so many of us learned in grade school Earth science courses: our planet's hydrosphere is a single, interconnected system.

This realization was the beginning of a new era in my work—one recognizing that it is this "compartmentalized" understanding of water that has led to so many of the poor management practices that we have as a society. It is our insistence on confining water policies and water rights into the autonomous circles of "agriculture," "energy," "industrial use," "human consumption," "rivers and streams," "oceans," and so on that has led to wholly independent and often entirely incongruent systems of standards, measurements, and practices. Confined to neat bubbles of discussion and management, we've failed to build and maintain intelligent infrastructure, and too often, in our pursuit of progress, we've completely destroyed the water-shaping ecosystems that could have provided sustainable solutions. It truly is time for us to redefine what it means to live on a water planet.

Called to this legacy, I have dedicated my life to continuing the environmental storytelling tradition of my father and grandfather. My work took a decidedly global scope when Blue Legacy, the nonprofit I founded in 2008, secured funding to launch our first expedition in February 2009. Traveling from the oceans that in so many ways have defined much of my family's history, I made my way to the roof of the world—to the Himalaya, where the tallest mountains on Earth rise and seven great rivers begin. Fed by the snow and ice blanketing these stunning peaks, the Ganges, Mekong, Yangtze, and their four sister rivers support a stunning two billion people—nearly one out of every three people in the world.

My expedition team and I pushed high into the mountains, tracing our way to the glaciers that feed those rivers. We stopped at the mountaintop temple of Kunjari, where tiger statues guard the gates and pilgrims ring bells for entrance. Pilgrims of a different sort, we climbed what felt like a million steps, pausing often to catch our breath in the thin mountain air. And at the top, we stood still for a moment, lifted our eyes, and gazed out at the landscape.

As stunning as it was, we had not traveled all these miles and climbed these steps to see its beauty—we were there to tell the story of water. And in the shadow of the mighty mountains, we witnessed fragility.

The Himalaya are the largest storehouse of fresh water outside the polar ice caps. And like so many others, the Himalayan glaciers are melting as a result of global climate change. They are retreating 30 to 50 feet every year, with science predicting that they could disappear altogether within 25 years—perhaps even sooner.

As Professor Veer Bhadra Mishra, the *mahant* (spiritual and administrative head) of Varanasi's second largest Hindu temple and an ardent environmentalist, would share with us so poignantly, the "waters of Mother Ganga [the Ganges] would someday flow in the Hudson," would fall on the plains of Africa, would make a cup of tea. And yet, for the nearly 1 billion Hindus of the world, 900 million of whom live in India, the seasonality brought to the Ganges by climate change would spell the end of the most enduring symbol of the world's oldest "living" religion. We strolled the streets and villages with that inescapable thought in mind. The sacred promise for so many was quickly being laid low by a century's quest for progress—the pollution clouding its pools was stealing health while failing sources were robbing hope.

The impact on the rivers and everything that depends upon them—from rainfall patterns to wildlife to agriculture—will be disastrous. Ecosystems will collapse. Hundreds of millions of people will be displaced. And the steady rivers that once shaped the course of society, and shaped the soul of cultures—will shift seasonally, delivering floods and landslides to some and unimaginable drought to others.

This is not theory. It is not opinion. It is a fact. Unless we do something, all of this will come to pass within our own generation.

Our 2009 expedition took us across five continents, as we worked to chronicle our planet's life-support system. Our destinations included Botswana's Okavango Delta, the Ganges River, the Dead Sea Basin, Cambodia, the Mississippi River, and Australia's western coast.

My grandfather used to tell me, "Go and see."

We wanted to go and see for ourselves what we understood in the abstract: the interconnectedness of water and the fragility of our most precious resource. In our inaugural expedition, we wanted to film and post from the field stories that illustrate how each of us experiences—how all of us experience—hopes for a better life and needs for life itself through this fragile gift we call water.

From outer space, it looks like we live on a blue planet. And we do—70 percent of Earth is covered by water. But only a tiny percentage of that water is fit for human consumption. If you could fit all the water in the world into a gallon jug, less than a teaspoon of that would be available for our use. And as glaciers on all continents retreat, that teaspoon is shrinking.

That isn't all. The quality of the water in the teaspoon is declining as well. Runoff and pollution have led to "dead zones" throughout our oceans. In India, hundreds of tanneries dump chemicals into the sacred Ganges. Aquifers are falling in the Middle East, in the United States, and around the world. A simple look at water consumption in the great and growing cities of the world makes it clear that we are racing down a collision course with history. Doomed to be remembered by the same epitaph that marks the ghostly ruins of the once great Khmer Empire at Angkor Wat: Unsustainable consumption of resources, mismanagement of critical infrastructure, and destruction of water-shaping ecosystems during a period of climate change lead to collapse.

The rulers of Angkor may not have understood what they were risking, but we do. Water—from droughts, storms, and floods to degrading water quality—is the defining challenge of our century. We'll see major conflicts over water and the proliferation of water refugees.

In the face of these issues, much of the world still discusses water-related policies in tiny buckets. We talk about river policy, fishing issues, farming standards, oceans, energy, and clean drinking water as though each exists in its own silo. But water does not recognize categories. And, as Israeli and Palestinian students working together to solve environmental issues in the Great Rift Valley shared with me, water does not recognize borders. It flows across boundaries connecting issues and people and positioning all of us, quite literally, downstream from each other.

Water is life. From spiritual leaders in India and government leaders in Africa to farmers in Jordan and Missouri, from Saudi royals to fishermen in Louisiana, all people say the same thing.

Water is life, they told me. Water is life.

We inhabit a water planet, and unless we protect, manage, and restore this precious resource, the future will be a very different place from the one we imagine today. The environmental movement is too often seen as a special interest group, but water is something we all have in common and it must, to defy that kind of labeling. It bridges our different walks of life. And its fragility is our fragility.

In this moment, on this issue, we all must see ourselves as environmentalists.

Our water crisis is a global issue—a human issue. It isn't going to be fixed by scientists or policymakers alone. It's up to each of us . . . to all of us.

When my grandfather followed his thirst to know more of what lay beneath the waves that called to him, he improvised, invented, and collaborated with trusted friends to craft the tools that would take him where few had ever gone and opened for the generations to follow a world we now better understand. My father followed in his footsteps—taking to the skies to capture a broader vision, and to the editing room to tell stories that spoke not just of the wonder for the new, but also of the weight of the known. The passion of his too short life set the course upon which I now find myself. And there are moments when it seems I hear his voice urging me onward—to ask the hard questions, to challenge old thinking, to carry forward a legacy that inspires some to innovate, others to discover, and each of us to act.

In the end, our shared humanity and our shared history are written in water. And how the story ends is entirely up to us.

HOW THIS BOOK CAME TO BE

AFTER A SCREENING of my documentary *Flow: For Love of Water,* Barbara Brownell Grogan and Susan Straight at National Geographic Books invited me to come to Washington to discuss a book project on water. I was extremely happy about this idea, because after spending five years against all odds making the documentary, I felt that although the film covers many subjects and offers some solutions from the perspective of inspiring people who have devoted their lives to water, the structure and time constraints of a feature-length movie give you only so much time to explore issues, whereas the written word could afford us a whole new arena in which to talk about complex ideas and experiences.

There are lots of innovative solutions, both potential and already in place; there are incredible people monitoring rivers, cleaning up watersheds, working day and night to find cost-effective ways to make water safe in places (way too many) where children die of waterborne diseases; there are groups tirelessly engaged in education and information programs; and the list goes on across the globe. But it's not very often that we get to hear their stories highlighted on prime-time news. It's so much more likely that we will hear about a water disaster, about human tragedies and lives lost, or get general information in a sound bite that falls far short of the whole story of a village or a community grappling with these problems. There is just so much to digest these days.

Storytelling, however, is in our genes, and stories have been told since the dawn of civilization. Grandmothers plant seeds of consciousness in small children's hearts; elders sing around a fire of their tribe's history. Stories are simultaneously new and old, crossing oceans of time, bridging generations and cultural differences, and reminding us of our communal inheritance and responsibility—the planet and its peoples. To quote renowned novelist Reynolds Price:

> A need to tell and hear stories is essential to the species *Homo sapiens*—second in necessity apparently after nourishment and before love and shelter. Millions survive without love or home, almost none in silence; the opposite of silence leads quickly to narrative, and the sound of story is the dominant sound of our lives, from the small accounts of our day's events to the vast incommunicable constructs of psychopaths.

At our first meeting at National Geographic, we agreed that there had already been many books covering various problems around the globe, but not that many about hope, cutting across theories and technicalities to tell simple, powerful, direct stories of passionate people and their involvement with water. We felt that giving a personal voice to water in the form of storytelling might bring people closer to home and give many readers a chance to connect more intimately to our subject.

Thanks to their incredible spirits of generosity, all the contributors participated joyfully in a common effort to share their experiences with us and to bring these stories to life. Whether on the mountaintop in the Himalaya or on the banks of a river in the United States or in a dusty village in Africa, all the essayists have in common a profound respect for and understanding of the natural world and water's role on our planet. They have all demonstrated a willingness to listen to the stories, ancient and new, of the Earth.

I feel that our love and respect for water must be restored. I hope that one of these stories, if not more, will percolate into your heart and mind, reminding you of this everlasting connection that we have with water.

—*Irena Salina*

AUTHORS' WEBSITES

Peter Gleick
Pacific Institute
www.pacinst.org

Frank Clifford
Frank Clifford articles
http://articles.latimes.com/writers/frank-clifford
http://articles.latimes.com/writers/frank-clifford/2

Marion Stoddart
Nashua River Watershed Association
www.nashuariverwatershed.org/index.html

The Work of 1000 (documentary)
www.workof1000.com

Anupam Mishra
Review of *The Radiant Raindrops of Rajasthan*
www.indiawaterportal.org/taxonomy/term/1074

Sandra Postel
Global Water Policy Project
www.globalwaterpolicy.org

Fred Pearce
Review of *Confessions of an Eco-Sinner*
www.greenprophet.com/2008/06/18/613/book-review-confessions-eco-sinner

Maude Barlow
The Council of Canadians
www.canadians.org

Blue Planet Project
www.blueplanetproject.net

Rose George
Rose George Website
http://rosegeorge.com

Shekhar Kapur
Shekhar Kapur Blog
www.shekharkapur.com/blog

Bill McKibben
Bill McKibben Website
www.billmckibben.com

Alton Byers
Himalaya—Changing Landscapes (photography exhibit)
www.changing-landscapes.com

The Mountain Institute
www.mountain.org

Melanie Stiassny
American Museum of Natural History—Melanie L. J. Stiassny Biography
www.research.amnh.org/ichthyology/staff/mljs/mljs.html

Ellsworth Havens
Water and Sanitation Rotarian Action Group
www.wasrag.org

AUTHORS' WEBSITES

Rajendra Singh
Tarun Bharat Sangh
www.tarunbharatsangh.org

William "Waterway" Marks
Water Voices From Around the World
www.watervoices.com

Scott Harrison
charity: water
www.charitywater.org

Lynne Cherry
Lynne Cherry Website
www.lynnecherry.com

Alex Matthiessen
Riverkeeper
www.riverkeeper.org

Christine Todd Whitman
The Whitman Strategy Group
www.whitmanstrategygroup.com/ourteamctw2.html

Jorge Recharte
The Mountain Institute
www.mountain.org

Frederick Kirschenmann
Leopold Center for Sustainable Agriculture
www.leopold.iastate.edu

Dave Rosgen
Wildland Hydrology
www.wildlandhydrology.com

Encyclopedia of World Biography—Dave Rosgen Biography
www.notablebiographies.com/newsmakers2/2005-Pu-Z/Rosgen-Dave.html

Kathy Robb
The Water Policy Institute
http://www.hunton.com/Resources/Sites/general.aspx?id=527

Women's Network for a Sustainable Future
www.wnsf.org

Ashok Gadgil
Lawrence Berkeley National Laboratory—Ashok Gadgil Biography
http://eetd.lbl.gov/staff/gadgil/AGadgil.html

WaterHealth International
www.waterhealth.com

James Workman
Blue Living Ideas
http://bluelivingideas.com/topics/saving-water/heart-of-dryness-a-new-book-and-interview-from-james-g-workman

Heart of Dryness
www.heartofdryness.com

Alexandra Cousteau
Blue Legacy, Expedition: Blue Planet
www.bluelegacy.net

GET INVOLVED

American Rivers
American Rivers, founded in 1973, protects and restores the United States' rivers for the benefit of people, wildlife, and nature. American Rivers has more than 65,000 members and supporters, with offices in Washington, D.C., and nationwide.

American Rivers
1101 14th St. N.W., Suite 1400
Washington, DC 20005
Phone: 202-347-7550
Fax: 202-347-9240
Website: www.americanrivers.org

Blue Planet Run
Blue Planet Run Foundation works with donors and project implementers to help raise and allocate sufficient funds to provide sustainable, locally owned water projects for those in need around the world.

Blue Planet Run
P.O. Box 3059
Redwood City, CA 94064-3059
Phone: 415-762-4340
Fax: 415-762-4340
Email: info@blueplanetrun.org
Website: http://blueplanetrun.org

charity: water
charity: water is a nonprofit organization bringing clean, safe drinking water to people in developing nations. It gives 100 percent of the money raised to direct project costs, funding sustainable clean water solutions in areas of greatest need.

200 Varick St., Suite 201
New York, NY 10014
Phone: 646-688-2323
Fax: 888-707-6466
Email: info@charitywater.org
Website: www.charitywater.org

Clean Water Action
Clean Water Action is an organization of 1.2 million members working to empower people to take action to protect America's waters.

Clean Water Action, DC Office
1010 Vermont Ave. N.W.
Suite 1100
Washington, DC 20005-4918
Phone: 202-895-0420
Fax: 202-895-0438
Website: www.cleanwateraction.org

EarthEcho International
EarthEcho International's programs are dedicated to the legacy of Philippe Cousteau, Sr., a legacy of devotion to the conservation and restoration of the oceans and one that can be shared by all people.

EarthEcho International
888 16th St. N.W., Suite 800
Washington, DC 20006
Phone: 202-349-9828
Fax: 202-355-1399
Email: info@earthecho.org
Website: www.earthecho.org

Earthwatch

For nearly 40 years, Earthwatch has brought together an inclusive community of scientists, corporate and nonprofit partners, teachers, students, and everyday citizens to find creative ways to respond to needs.

Earthwatch
3 Clock Tower Place, Suite 100
Box 75
Maynard, MA 01754
Phone: 800-776-0188
Fax: 978-461-2332
Email: Expeditions@earthwatch.org
Website: www.earthwatch.org

Ecotact Limited, Innovative Sanitation

Ecotact is a social enterprise incorporated in Kenya in 2006 with the sole objective of developing innovative social investments in environmental sanitation and management in Africa and beyond. Ecotact Limited was founded by David Kuria, a Kenyan Ashoka Fellow.

Ecotact Limited
Menelik Road, Off Ngong Road
P.O. Box 24045
00100 GPO
Nairobi, Kenya
Phone: 254-20-2459130
Cellphone: 254-772751581
Email: office@ecotact.org
Website: www.ecotact.org/ecotact/index.php

Engineers Without Borders-USA (EWB-USA)

EWB-USA helps create a more stable and prosperous world by addressing people's basic human needs by providing necessities such as clean water, power, sanitation and education.

Engineers Without Borders-USA
4665 Nautilus Court, Suite 300
Boulder, CO, 80301
Phone: 303-772-2723
Fax: 303-772-2699
Website: www.ewb-usa.org

Food & Water Watch

Food & Water Watch is a nonprofit organization that works to guarantee safe, wholesome food produced in a humane and sustainable manner and public, rather than private, control of water resources including oceans, rivers, and groundwater.

Food & Water Watch, D.C. Office
1616 P St. N.W.,
Suite 300
Washington, DC 20036
Phone: 202-683-2500
Fax: 202-683-2501
Website: www.foodandwaterwatch.org

International Rivers

International Rivers, founded in 1985, works toward a world where everyone has access to clean water and energy, and where development projects neither degrade nature nor destroy communities. Work is focused in Latin America, Asia, and Africa.

International Rivers
2150 Allston Way
Suite 300
Berkeley, CA 94704-1378
Phone: 510-848-1155
Fax: 510-848-1008
Email: info@internationalrivers.org
Website: http://internationalrivers.org

International Water and Sanitation Centre (IRC)

IRC works with partners for an improved, low-cost water supply, sanitation, and hygiene in developing countries.

IRC
P.O. Box 82327
2508 EH
The Hague, The Netherlands
Phone: 31 70 3044000
Fax: 31 70 3044044
Website: www.irc.nl

Ji Duma

Ji Duma helps build low-cost, community-based water projects in Mali and works in a collaborative manner with villagers to improve the quality of their lives.

Practical Small Projects
c/o Kaufcan & Canoles
P.O. Box 6000
Williamsburg, VA 23188

Loriana Dembélé (President)
BP 1024
Bamako, Mali
Email: @jiduma.org
Website: www.jiduma.org

Natural Resources Defense Council (NRDC)

NRDC works on a broad range of issues as it pursues its mission to safeguard the Earth, its people, its plants and animals, and the natural systems on which all life depends.

Natural Resources Defense Council
40 West 20th St.
New York, NY 10011
Phone: 212-727-2700
Fax: 212-727-1773
Website: www.nrdc.org/issues

Oxfam America

Oxfam America, an affiliate of Oxfam International, is an international relief and development organization that creates lasting solutions to poverty, hunger, and injustice. Together with individuals and local groups in more than 100 countries, Oxfam saves lives, helps people overcome poverty, and fights for social justice.

Oxfam America
226 Causeway St., Fifth Floor
Boston, MA 02114-2206
Phone: 800-776-9326
Fax: 617-728-2594

Email: action@oxfamamerica.org
Website: www.oxfam.org

Playpumps International
PlayPumps International's mission is to help improve the lives of children and their families by providing easy access to clean drinking water, enhancing public health, and offering play equipment to millions across Africa. The organization provides schools and communities throughout Africa with an innovative, sustainable, free supply of clean drinking water.

Website: www.playpumps.org

Potters for Peace
Potters for Peace is a U.S.-based nonprofit network of potters, educators, technicians, supporters, and volunteers. Founded in Nicaragua in 1986, the group works primarily in Central America, although its water filter projects are worldwide.

Potters for Peace
P.O. Box 1043
Bisbee, AZ 85603
Phone: 520-249-8093
Email: peterpfp@gmail.com (Peter Chartrand)
Website: www.pottersforpeace.org

Project WET: Worldwide Water Education
Since 1984, Project WET a nonprofit organization, has dedicated itself to the mission of reaching children, parents, teachers, and community members of the world with water education.

Project WET Foundation
1001 West Oak St., Suite 210
Bozeman, MT 59715
Phone: 406-585-2236 , 866-337-5486 (toll free in the U.S.)
Fax: 406-522-0394
Email: info@projectwet.org
Website: www.projectwet.org

Sierra Club
America's oldest, largest, and most influential grassroots environmental organization working to protect communities and the planet.

Sierra Club
85 Second St., Second Floor
San Francisco, CA 94105
Phone: 415-977-5500
Fax: 415-977-5799
Email: information@sierraclub.org
Website: www.sierraclub.org

Silent Spring Institute
Silent Spring Institute builds on a unique partnership of scientists, physicians, public health advocates, and community activists to identify and break the links between the environment and women's health problems, especially breast cancer.

Silent Spring Institute
29 Crafts St.
Newton, MA 02458
Phone: 617-332-4288
Fax: 617-332-4284

Email: info@silentspring.org
Website: www.silentspring.org/

UNICEF

UNICEF works in more than 90 countries around the world to improve water supplies and sanitation facilities in schools and communities, and to promote safe hygiene practices.

U.S. Fund for UNICEF
125 Maiden Lane, 11th Floor
New York, NY 10038
Phone: 212-686-5222
Website: www.unicef.org/wash

The United Nations Educational, Scientific, and Cultural Organization (UNESCO) World Heritage Centre

UNESCO seeks to encourage the identification, protection and preservation of cultural and natural heritage around the world considered to be of outstanding value to humanity.

World Heritage Centre
7 place de Fontenoy
75352 Paris 07 SP France
Phone: 33 0 1 45 68 15 71
Fax: 33 0 1 45 68 55 70
Email: wh-info@unesco.org
Website: http://whc.unesco.org

WASH in Schools

WASH in Schools in an International Water and Sanitation Centre. One of its goals is to teach schoolchildren the practices of basic hygiene at school.

WASH in Schools
P.O. Box 2869
2601 CW Delft
The Netherlands
Phone: 31 15 219 2939
Fax: 31 15 219 2973
Website: www.schools.watsan.net/

Water Advocates

Water Advocates is the first U.S.-based nonprofit organization dedicated solely to increasing American support for worldwide access to safe, affordable, and sustainable supplies of drinking water and adequate sanitation.

Water Advocates
1506 21st St. N.W.
Washington DC 20036
Phone: 202-293-4002
Fax: 202-293-4001
Email: info@wateradvocates.org
Website: www.wateradvocates.org

Waterkeeper Alliance

Founded in 1999 by environmental attorney and activist Robert F. Kennedy, Jr., and several veteran Waterkeepers, Waterkeeper Alliance is a global movement of on-the-water advocates who patrol and protect more than 100,000 miles of rivers, streams, and coastlines in North and South America, Europe, Australia, Asia, and Africa.

Waterkeeper Alliance
50 South Buckhout St., Suite 302
Irvington, NY 10533
Phone: 914-674-0622
Email: info1@waterkeeper.org
Website: www.waterkeeper.org

Water.org

Co-founded by Matt Damon and Gary White, Water.org is a non-profit organization that has transformed hundreds of communities in Africa, South Asia, and Central America by providing access to safe water and sanitation. In July 2009, WaterPartners merged with H2O Africa, resulting in the launch of Water.org. Water.org works with local partners to deliver innovative solutions for long-term success. Its microfinance-based WaterCredit Initiative is pioneering sustainable giving in the sector.

Water.org
920 Main St., Suite 1800
Kansas City, MO 64105
Phone: 816-877-8400
Website: http://water.org

Water & Sanitation Rotarian Action Group

Rotarian Action Group provides information, support, and encouragement to Rotarians, Rotary clubs, and districts to take active roles in projects/programs to develop safe water and sanitation as a means of promoting health and alleviating hunger.

Rotarian Action Group for Population Growth and Sustainable Development
344 West Pike St.
Lawrenceville, GA 30045
Phone: 770-407-5633
Website: www.wasrag.org

Wherever the Need

Wherever the Need is a U.S. nonprofit focusing on a holistic approach to the provision and use of eco-sanitation toilets and water.

Wherever the Need
P.O. Box 13056
Marina del Rey, CA 90292
Phone: 310-216-7895
Website: www.wherevertheneed.org

Worldwatch Institute

The Worldwatch Institute is an independent research organization that provides fact-based analysis of critical global issues. Its mission is to generate and promote insights and ideas that empower decision-makers to build an ecologically sustainable society that meets human needs.

Worldwatch Institute
1776 Massachusetts Ave. N.W.
Washington, DC 20036-1904
Phone: 202-452-1999
Fax: 202-296-7365
Email: worldwatch@worldwatch.org
Website: www.worldwatch.org/

FURTHER READING

Altman, Nathaniel. *Sacred Water: The Spiritual Source of Life.* Mahwah, New Jersey: HiddenSpring, 2002.

Balaguer, Alejandro, Alan Garcia Perez, and Jorge Recharte. *Agua Madre/Mother Water: Truths and Images.* Lima, Peru: Graph Ediciones.

Barlow, Maude. *Blue Covenant: The Global Water Crisis and the Coming Battle for the Right to Water.* New York: New Press, 2008.

Baviskar, Amita. *Waterlines. The Penguin Book of River Writings.* New Delhi: Penguin India, 2003.

Brown, Lester R. *Plan B 4.0: Mobilizing to Save Civilization.* New York: W. W. Norton and Company, 2009.

Carson, Rachel. *Silent Spring.* Boston: Houghton Mifflin, 1962.

Cherry, Lynne, and Gary Braasch. *How We Know What We Know About Our Changing Climate: Scientists and Kids Explore Global Warming.* Nevada City, California: Dawn Publications, 2008.

Childs, Craig. *The Secret Knowledge of Water: Discovering the Essence of the American Desert.* Seattle: Sasquatch Books, 2000.

Clifford, Frank. *The Backbone of the World: A Portrait of the Vanishing West along the Continental Divide.* New York: Broadway Books, 2003.

Coats, Callum. *Living Energies: An Exposition of Concepts Related to the Theories of Viktor Schauberger.* Bath, United Kingdom: Gateway Books, 2002.

Emoto, Masaru. *The Hidden Messages in Water.* New York: Atria Books 2005.

George, Rose. *The Big Necessity: The Unmentionable World of Human Waste and Why It Matters.* New York: Metropolitan Books, 2008.

Gleick, Peter. *The World's Water 2008-2009: The Biennial Report on Freshwater Resources.* Washington, D.C.: Island Press, 2008.

Glennon, Robert. *Water Follies: Groundwater Pumping and the Fate of America's Fresh Waters.* Washington, D.C.: Island Press, 2002.

Hollyer, Beatrice. *Our World of Water.* New York: Henry Holt and Company, 2009.

Hornbein, Thomas F. *Everest: The West Ridge.* Seattle: Mountaineers Books, 1964.

Jehl, Douglas, and Bernadette Mcdonald, eds. *Whose Water Is It? The Unquenchable Thirst of a Water-Hungry World.* National Geographic, 2003.

FURTHER READING

Lancaster, Brad. *Rainwater Harvesting for Drylands and Beyond,* Vol. 1. White River Junction, Vermont: Chelsea Green Publishing Company, 2006.

———. *Rainwater Harvesting for Drylands and Beyond,* Vol. 2. Tucson, Arizona: Rainsource Press, 2007.

Linden, Eugene. *The Wind of Change: Climate, Weather, and the Destruction of Civilizations.* New York: Simon and Schuster, 2007.

Marks, William E., ed. *Water Voices From Around the World.* Self-published, 2008.

McCully, Patrick. *Silenced Rivers: The Ecology and Politics of Large Dams.* New Delhi: Orient Longman, 1998.

McDonough, William, and Michael Braungart. *Cradle to Cradle.* New York: North Point Press, 2002.

McKibben, Bill. *The End of Nature.* New York: Random House, 2006.

Midkiff, Ken, and Robert F. Kennedy, Jr. *Not a Drop to Drink: America's Water Crisis (and What You Can Do).* Novato, California: New World Library, 2007.

Mishra, Anupam. *The Radiant Raindrops of Rajasthan.* New Delhi: Research Foundation for Science, Technology, and Ecology, 2001.

Patel, Raj. *Stuffed and Starved: The Hidden Battle for the World Food System.* New York: Melville House, 2008.

Pearce, Fred. *Confessions of an Eco Sinner: Tracking Down the Sources of My Stuff.* London: Eden Project Books, 2008.

Pollan, Michael. *The Omnivore's Dilemma: A Natural History of Four Meals.* New York: Penguin, 2007.

Postel, Sandra, and Brian Richter. *Rivers for Life: Managing Water for People and Nature.* Washington, D.C.: Island Press, 2003.

Powell, J. W. *The Exploration of the Colorado River and Its Canyons.* Mineola, New York: Dover Publications, 1961.

Powell, James Lawrence. *Dead Pool: Lake Powell, Global Warming, and the Future of Water in the West.* Berkeley: University of California Press, 2009.

Reisner, Marc. *Cadillac Desert: The American West and Its Disappearing Water.* New York: Penguin, 2003.

Roddick, Anita, with Brooke Shelby Biggs. *Troubled Water: Saints, Sinners, Truths and Lies About the Global Water Crisis.* White River Junction, Vermont: Chelsea Green Publishing Company, 2004.

Royte, Elizabeth. *Bottlemania: How Water Went on Sale and Why We Bought It.* New York: Bloomsbury USA, 2008.

Schmidt, Gavin, and Joshua Wolfe. *Climate Change: Picturing the Science.* New York: W. W. Norton and Company, 2009.

Shnayerson, Michael. *Coal River.* New York: Farrar, Straus and Giroux 2008.

Specter, Michael. "The Last Drop: Confronting the possibility of a global catastrophe." *New Yorker* (October 23, 2006). Available online at http://www.newyorker.com/archive/2006/10/23/061023fa_fact1.

Speth, James Gustave. *Red Sky at Morning: America and the Crisis of Global Environment.* New Haven, Connecticut: Yale University Press, 2004.

de Villiers, Marq. *Water: The Fate of Our Most Precious Resource.* New York: Mariner Books, 2001.

"Water: A Global Challenge." *Journal of International Affairs* (Spring/Summer 2008).

Waterman, Jonathan. *Running Dry: A Journey From Source to Sea Down the Colorado River.* Washington, D.C.: National Geographic, 2010.

Wood, Chris. *Dry Spring: The Coming Water Crisis of North America.* Vancouver, British Columbia: Raincoast Books.

Workman, James G. *Heart of Dryness: How the Last Bushmen Can Help Us Endure the Coming Age of Permanent Drought.* New York: Walker and Company, 2009.

News Websites

National Geographic:
Freshwater Crisis
www.nationalgeographic.com/freshwater/

Circle of Blue WaterNews
www.circleofblue.org/waternews/

Rainwater Harvesting-USA
www.ecobusinesslinks.com/rainwater-harvesting.htm

U.S. Drought Monitor Maps
http://drought.unl.edu/dm/monitor.html

The Great Garbage Patch
www.greatgarbagepatch.org/